Introduction to **Carbon Science**

Introduction to **Carbon Science**

Ian A. S. Edwards
Harry Marsh
Rosa Menendez
Brian Rand
Sebastian West
Andrew J. Hosty
Khim Kuo
Brian McEnaney
Timothy Mays
David J. Johnson
John W. Patrick
David E. Clarke
Jack C. Crelling
Ralph J. Gray

Editor **Professor Harry Marsh**
University of Newcastle upon Tyne, U.K.

Butterworths
London Boston Singapore Sydney Toronto Wellington

PART OF REED INTERNATIONAL P.L.C.

All rights reserved. No part of this publication may be reproduced in any material form (including photocopying or storing it in any medium by electronic means and whether or not transiently or incidentally to some other use of this publication) without the written permission of the copyright owner except in accordance with the provisions of the Copyright, Designs and Patents Act 1988 or under the terms of a licence issued by the Copyright Licensing Agency Ltd, 33–34 Alfred Place, London, England WC1E 7DP. Applications for the copyright owner's written permission to reproduce any part of this publication should be addressed to the Publishers.

Warning: The doing of an unauthorised act in relation to a copyright work may result in both a civil claim for damages and criminal prosecution.

This book is sold subject to the Standard Conditions of Sale of Net Books and may not be re-sold in the UK below the net price given by the Publishers in their current price list.

First published 1989

© Butterworth & Co. (Publishers) Ltd, 1989

British Library Cataloguing in Publication Data

Introduction to carbon science
  1. Carbon
  I. Marsh, Harry
  546′.6811

ISBN 0-408-03837-3

Library of Congress Cataloging-in-Publication Data

Introduction to carbon science/edited by Harry Marsh.
    p. cm.
  Includes bibliographies.
  ISBN 0-408-03837-3 :
  1.Carbon.   I. Marsh, Harry.
  QD181.C1I48   1989
  546′.68–dc 19   89-721

Photoset by Scribe Design, Gillingham, Kent
Printed and bound by Hartnoll Ltd, Bodmin, Cornwall

Foreword

The international community of carbon scientists, technologists and engineers meet annually and in 1988 the venue was the University of Newcastle upon Tyne, U.K.

Large international meetings usually follow the same format of plenary lectures, short presentations of studies and poster sessions. To those entering the field of carbon science for the first time, either as young people or those who are changing direction of work as part of their career, such a scientific meeting can be somewhat bewildering.

The idea of a Short Course as an introductory series of lectures prior to the main Conference is not new in itself but is new to the carbon community. This book follows the sequence of topics of such a Short Course, each Chapter being built upon the material used by the lecturer in the Short Course of September 1988.

Carbon Science, as a general term, can be very far-reaching, ranging from polymer science to prosthetics, from crystallography to carbonization, spectroscopy to surface science, etc.. Not all aspects could possibly be covered in one week of lectures, *i.e.* nine sessions.

The planning committee considered that the following topics represented the essentials of a foundation course in carbon science:-

1. Structure in carbons and carbon forms.
2. Mechanisms of formation of isotropic and anisotropic carbons.
3. Physical properties of pitch relevant to the fabrication of carbon materials.
4. Kinetics and catalysis of carbon gasification.
5. Porosity in carbons and graphites.
6. Carbon fibres: manufacture, properties, structure and applications.
7. Mechanical properties of cokes and composites.
8. The nature of coal material.
9. Coal to coke transformations.

The intention that the Short Course material should be collected in a permanent and accessible form has been realised.

This book is an introduction to carbon science, it is contemporary and it is anticipated that it will be of interest both to the beginner and to the experienced. For the reader requiring a wider knowledge of specific topics, the text is supplemented with appropriate and extensive references. However, the text alone gives full grounding of the subject.

Harry Marsh

Acknowledgements

This book is a collaborative effort, not only in producing Chapters of related topics, but in taking care to prevent overlap and with limited, but necessary, cross-referencing.

Those who have expertise in their subjects have heavy work-loads these days and I am indeed honoured to have had the generous support of the authors of the Chapters who are distinguished in their subject areas and who have taken such care with their manuscripts.

The final texts have been put together by Bridget A. Clow, Marion Poad and Elaine Watson. Assistance with diagrams has been most kindly given by Eduardo Romero Palazon. Brenda Chan supported me in the last stages of compilation. Ed Heintz (University of Buffalo, U.S.A.) was a tremendous help as co-ordinator during the running of the short course.

Finally, the professional assistance of the office of Butterworths, from Jayne Holder in particular, is appreciated.

Harry Marsh

An Introduction to Authors

Professor J.C. Crelling
Department of Geology, Southern Illinois University
Carbondale, U.S.A.

Jack C. Crelling graduated in geology from the University of Delaware, U.S.A. After his Ph.D. studies at Penn. State University, he spent five years in the Coal Research Laboratory of Bethlehem Steel Corporation before joining Southern Illinois University where he is presently Professor of Geology and the Head of the Coal Characterisation and Maceral Separation Laboratories. His area of research is coal petrology, including fluorescence spectro-analysis and photo-acoustic microscopy. His most recent work involves the separation of single coal macerals and the determination of their petrology, chemistry and reactivity. He is Chairman of the Coal Division of the Geological Society of America, President of the Society for Organic Petrology and serves on the Editorial Board of Fuel Processing Technology.

Dr. D.E. Clarke
Carbon Research Group
Loughborough Consultants Ltd.
Loughborough University of Technology, U.K.

David E. Clarke graduated from the University of Bath in 1978 and entered journalism as an editorial staff member of the Journal FUEL, working under the guidance of the then Editor-in-Chief, Dr. I.C.G. Dryden. In 1982, he joined the Northern Carbon Research Laboratories, University of Newcastle upon Tyne to carry out research under the direction of Professor H. Marsh on the coal-to-coke transformation, briquetting and also the degradation of metallurgical cokes within the blast furnace and was awarded a Ph.D. in 1986. Since 1987, he has been working with Dr. J.W. Patrick at Loughborough on coal liquefaction and continuing studies on metallurgical coke degradation. He is currently a member of the RSC/IOP Joint Carbon Group Committee and Hon. Editor of the CARBON NEWSLETTER.

Dr. I.A.S. Edwards
Northern Carbon Research Laboratories
University of Newcastle upon Tyne, U.K.

Ian A.S. Edwards studied chemistry at King's College, University of Durham, his postgraduate research being in the field of crystallography. The subject of his Ph.D. thesis was 'Crystal Structure Determinations by X-ray Methods'. Following post doctoral research in surface analysis using LEED and AES in the Physical Chemistry Department of the University of Newcastle upon Tyne, he joined the Northern Carbon Research Laboratories in 1980. During the past eight years, he has supervised research in structural and reactivity measurements on a wide range of carbon materials including graphite, coke and pitch. He has presented papers at international conferences on crystallography and carbon science. He is Honorary Secretary of the local section of the Royal Society of Chemistry and Chairman of the local section of the Society of Chemical Industry.

Mr. R.J. Gray
Process and Energy Management Corporation
Pittsburgh, PA, U.S.A.

Ralph J. Gray is a coal scientist with 38 years experience in coal exploration, evaluation and development, his B.S. and M.S. being in geology. In 1957, after joining U.S. Steel, he began to develop a petrographic classification of coals which introduced vitrinite reflectance as V-Types and the first U.S. coke-strength prediction system based on coal petrography. He developed microscopic techniques for analysis of coke carbon forms, non-maceral microstructures and quinoline insolubles from coal-tar and pitches and automatic microscopic systems for monitoring coals and coal blend proportions for coke making. He has written 54 publications and received the Iron and Steel Societies Joseph Barker Award in 1986 for his contributions to coke making. In 1987, he worked with the EEC on the new international classification of coal. He retired from US Steel in 1983 and is currently with the Process and Energy Management Corporation, Pittsburgh, U.S.A.

Mr. A.J. Hosty and Mr. S.West
Division of Ceramics, Glasses & Polymers,
University of Sheffield, U.K.

Mr. Andrew J. Hosty and Mr. Sebastian West are graduate students in The School of Materials at The University of Sheffield. They both graduated with honours in Materials Science in 1987. Mr. Hosty is investigating the fabrication of carbon-carbon composites using carbonaceous mesophase as a matrix precursor and Mr. West is investigating rheological aspects of pitch-coal interactions.

Dr. D.J. Johnson
Department of Textile Industries
University of Leeds, U.K.

David J. Johnson is Reader of Textile Physics at the University of Leeds, and has worked on the structure/property relationships of carbon fibres since 1968. He has published extensively and has contributed plenary lectures at a number of International Carbon Conferences. He collaborated with W. Watt and W. Johnson, the inventors of PAN-based carbon fibres, for many years and continues to carry out project work for industry. His main fields of experience are X-ray diffraction and electron microscopy, particularly lattice-fringe imaging; current research work is concerned with the compressive failure of carbon and other high-performance fibres. Dr. Johnson is Education Secretary of the Royal Microscopical Society.

Dr. A.K. Kuo
Northern Carbon Research Laboratories
University of Newcastle upon Tyne, U.K.

Khim Kuo graduated in chemistry from the University of Newcastle upon Tyne in 1982 and subsequently spent one year teaching and carrying out research in radiation chemistry at the University of Delaware, U.S.A. Following her return to the U.K. and M.Sc. studies in biochemistry, she joined the Northern Carbon Research Laboratories studying under the supervision of Professor H. Marsh. The subject of her Ph.D. studies was "The Concept of Active Surface Area in Carbon/Coke Gasification". At the time of publication, Dr. Kuo is working in the field of aluminium reduction cell lining materials at the Comalco Research Centre, Thomastown, Australia.

Professor H. Marsh
Northern Carbon Research Laboratories
University of Newcastle upon Tyne, U.K.

Harry Marsh graduated from The University of Durham. He studied under Professor H. L. Riley for a Ph.D. in coal chemistry, subsequently taking a D.Sc. in carbon science at The University of Newcastle upon Tyne. His professional career has centered around carbon and coal science with an emphasis on relating research to industrial requirements. Professor Marsh has published about 300 papers and conference proceedings and trained about 60 research students and 60 postgraduate researchers. He has lectured in all five continents. In 1985 he received the George Skakel Memorial Award from the American Carbon Society and, in 1989, the Henry H. Storch Award from the American Chemical Society for promoting interactions between Academe and industry. In 1989, he established a Coal Research Forum in the U.K. to promote both the interests of coal within the U.K. and the influence on coal markets abroad. Currently, he holds a Chair in Carbon Science and is Head of the Northern Carbon Research Laboratories. He is a Principal Editor of FUEL, on the Editorial Board of CARBON, Chairman of the Industrial Carbon and Graphite Group of the Society of Chemical Industry and is a member of the sub-committee of SERC with interests in coal.

Dr. T.J. Mays
School of Materials Science
University of Bath U.K.

Tim Mays is a Research Fellow in the School of Materials Science of the University of Bath. He graduated with a first-class honours degree in physics and, after working as a statistician with the CEGB, studied under Dr. B. McEnaney for his Ph.D. on gaseous diffusion and pore structure in graphites. He has recently worked in the Carbon Research Group in the School on the determination of pore structure in carbons using computer-aided quantitative microscopy and is currently researching into the fundamental theory of adsorption in microporous carbons. Dr. Mays is a member of the Committee of the Joint Carbon Group of the Institute of Physics and the Royal Society of Chemistry and recently co-edited, with Dr. McEnaney, the Proceedings of the International Carbon Conference, "Carbon 88".

Dr. B. McEnaney
School of Materials Science
University of Bath, U.K.

Brian McEnaney is a Reader and Head of the Carbon Research Group in the School of Materials Science of the University of Bath. He is presently Chairman of the Committee of the Joint Carbon Group of the Institute of Physics and the Royal Society of Chemistry and a member of the Editorial Board of the Journal CARBON. He has published over 100 papers mainly in the field of carbons and graphites but also on metallic corrosion. Recently, he co-edited with Dr. T.J. Mays the Proceedings of the International Carbon Conference, "Carbon 88". Dr. McEnaney presented Plenary Review Lectures on aspects of porosity in carbons and graphites at the 1985 American Carbon Conference and the 1986 German Carbon Conference.

Dr. R.M. Menendez
Instituto Nacional del Carbon,
Oviedo, Spain.

Rosa Menendez is a graduate of the University of Oviedo where, after her first degree in organic chemistry, she gained a masters degree in pollution control and, in 1986, a Ph.D. in the characterization of coal products. She spent two years (1987 - 1988) working in the Northern Carbon Research Laboratories, University of Newcastle upon Tyne, researching into coal and pitch carbonization. Currently, her work at INCAR is related to industrial applications of petrology, particularly carbonization and combustion. She is a member of the ICCP and the American Chemical Society.

Dr. J.W. Patrick
Carbon Research Group
Loughborough Consultants Ltd.
Loughborough University of Technology, U.K.

John W. Patrick obtained his B.Sc. in chemistry and his Ph.D. from the University of London for a study of carbon-sulphur complexes. He is currently Director of the Carbon Research Group at Loughborough University of Technology following 24 years service with the British Carbonization Research Association where he held the position of Head of Fundamental Studies. He has over 30 years experience in studying various aspects of carbon science ranging from the carbonization process to the gasification of carbon and the inter-relation of the strength and structure of cokes and carbons. He has made a special study of the strength of metallurgical cokes and its relation to the coke porous structure and to the nature of the coke carbon, i.e., optical texture. He is a Principal Editor of the Journal FUEL and a former chairman of the Joint Carbon Group of the Royal Society of Chemistry and the Institute of Physics.

Dr. B. Rand
Division of Ceramics, Glasses and Polymers
University of Sheffield, U.K.

Brian Rand graduated from King's College, University of Durham in chemistry taking his Ph.D. in carbon science under the supervision of Professor H. Marsh. After a period of industrial employment, Dr. Rand is currently Senior Lecturer in the School of Materials, University of Sheffield, U.K. He has published extensively with particular interest in thixotropic changes involved in reheating coal-tar pitches and the general rheological behaviour of pitch precursors used in composites and fibres. Emphasis is given to the significance of glass transition temperature. Other studies include porosity in carbons and surface characteristics of PAN fibres.

Contents

Foreword v

Acknowledgements vii

An Introduction to Authors ix

    J.C. Crelling ix
    D.E. Clarke ix
    I.A.S. Edwards x
    R.J. Gray x
    A.J. Hosty and S. West xi
    D.J. Johnson xi
    A.K. Kuo xi
    H. Marsh xii
    T.J. Mays xii
    B. McEnaney xiii
    R.M. Menendez xiii
    J.W. Patrick xiii
    B. Rand xiv

Chapter 1

Structure in Carbons and Carbon Forms

	Summary	1
1	**Introduction - Setting the Scene**	2
	1.1 The element carbon	2
	1.2 Bonding in carbon materials	3
	1.3 Diamond and graphite - perfect structures	4
2	**Order/Disorder**	6
	2.1 More-ordered structures	6
	2.2 Less-ordered structures	8
	2.3 Range of order	9
3	**Carbon Forms**	10
	3.1 Graphitic carbons (Natural and synthetic graphites)	10
	3.2 Non-graphitic carbons and graphitization	11
	3.3 Graphitizable and non-graphitizable carbons	11
	3.4 Pitches	11
	3.5 Cokes	12
	3.6 Coals	13
	3.7 Carbon fibres	14
	3.8 Other carbon materials	15
4	**Composites**	16
	4.1 Graphitic composites	17
	4.2 Carbon electrodes	17
	4.3 Carbon/carbon composites	18
5	**Methods of Studying Carbon Structure**	19
	5.1 Optical microscopy	19
	5.2 Electron microscopy (SEM and TEM)	22
	5.3 X-ray diffraction	27
	5.4 Raman spectroscopy	29
	5.5 Surface techniques	29
6	**Factors in Carbon Structures**	30

7	**Conclusions**		30
	7.1	The diversity of carbon	31
	References		31

Chapter 2

Mechanisms of Formation of Isotropic and Anisotropic Carbons

	Summary		37
1	**Introduction**		38
2	**Isotropic Carbon**		39
3	**Graphitizable Carbon - The problem**		46
	3.1	Background	46
	3.2	Mesophase: Early recognition	47
	3.3	Nematic liquid crystals	48
	3.4	Structure in liquid crystals	48
	3.5	Nucleation of mesophase	50
	3.6	Structure within mesophase	51
4	**Chemistry and Viscosity of Pyrolysis Systems**		55
	4.1	Growth and properties of mesophase: Summary	59
	4.2	Aspects of mesophase chemistry	59
	4.3	Mesophase growth and coalescence	60
	4.4	Carbons/cokes from mesophase from pitch	61
5	**Mesophase from Coal**		65
	5.1	Metallurgical coke	65
	5.2	Coal chemistry	65
	5.3	Mesophase formation during coal pyrolysis	66
	References		70

Chapter 3

Physical Properties of Pitch Relevant to the Fabrication of Carbon Materials

		Summary	75
1		**Introduction**	76
2		**Origins and Composition of Pitch**	77
	2.1	Coal-tar pitch	77
	2.2	Petroleum pitch	80
	2.3	Solubility as a characterisation technique	81
	2.4	Chemical characteristics	82
	2.5	Mesogenic character of pitch	82
3		**Structure of Pitch**	83
	3.1	Pitch as a glassy solid	83
	3.2	Pitch as a colloidal system	84
	3.3	Particulate inclusions	85
4		**Rheological Properties of Pitch**	86
	4.1	Newtonian and non-Newtonian flow	86
	4.2	Effect of temperature on the viscosity coefficient	88
	4.3	Measurement of the glass transition temperature	88
	4.4	Factors determining the glass transition temperature and other reference temperatures	89
	4.5	Effect of particulate matter on rheology	91
	4.6	Mesophase rheology	91
5		**Pyrolysis of Pitch**	92
	5.1	Transformation diagrams	93
	5.2	Uses of the transformation diagram	96
	5.3	Experimental diagram	96
6		**Pitch as a Binder and Matrix Material in Engineering Materials**	96
	6.1	Effect on porosity	96
	6.2	Surface activity of pitch	100

| 7 | Electrical Conductivity | 102 |

    References   103

Chapter 4

Kinetics and Catalysis of Carbon Gasfication

	Summary	107
1	Introduction	108
2	The Nature of Carbon Surfaces	108
3	Reactivity of Carbon	109
	3.1 Selective gasification	109
4	Reaction Kinetics and Mechanisms	112
	4.1 Chemical and diffusion control of rate	112
	4.2 Reaction rates	116
	4.3 Chemisorption and desorption	116
	4.4 Importance of active surface area (ASA) to reactivity	120
	4.5 Concept of reactivity	121
5	The Carbon-Molecular Oxygen Reaction	125
6	The Carbon-Carbon Dioxide Reaction	128
7	The Carbon-Steam Reaction	130
8	The Carbon-Oxides of Nitrogen Reaction	132
9	The Carbon-Hydrogen Reaction	132
10	Comparison of Carbon Gasification Reactions	133
11	Catalysis of Oxidation Reactions	134
	11.1 Effects of catalysts on reaction kinetics	134
	11.2 Mechanisms of catalysis	135

	11.3	Understanding of catalysis by oxygen-transfer reactions	136
	11.4	Topography of catalytic gasification	138
12	**Inhibition of the Gas-Carbon Reaction**		142
	References		145

Chapter 5

Porosity in Carbons and Graphites

	Summary		153
1	**Introduction**		154
	1.1	Classifications of porosity	154
	1.2	Some examples of porosity in carbons and graphites	156
2	**Effects of Porosity on Properties of Carbons**		160
3	**Densities of Carbons**		161
4	**Surface Areas from Gas Adsorption**		162
	4.1	Experimental methods	163
	4.2	The Brunauer-Emmett-Teller (BET) theory	164
	4.3	Fractal surfaces of carbons	166
5	**Surface Areas from Small Angle X-ray Scattering**		167
	5.1	The Debye equation	168
	5.2	The Porod law	170
	5.3	The Guinier equation	172
6	**Microporous Carbons**		172
	6.1	Adsorption in microporous carbons	173
	6.2	Calculations of adsorption potentials	174
	6.3	Application of the BET equation to microporous carbons	175
	6.4	The Dubinin-Radushkevich (DR) equation	177
	6.5	Estimations of the dimensions of micropores	180

7	**Mesoporous Carbons**	181
	7.1 The Kelvin equation	182
	7.2 Limitations of the Kelvin equation	184
8	**Macroporous Carbons**	185
	8.1 Mercury porosimetry	185
	8.2 Fluid transport in pores	188
	8.3 Image analysis	191
	References	194

Chapter 6

Carbon Fibres: Manufacture, Properties, Structure and Applications

	Summary	197
1	**Introduction**	198
	1.1 History	198
	1.2 General properties	200
2	**Preparation**	200
	2.1 Carbon fibres from PAN	200
	2.2 Carbon fibres from mesophase pitch	201
3	**Tensile Properties**	202
	3.1 Tensile modulus	202
	3.2 Tensile strength	202
	3.3 Practical properties of carbon fibres	203
4	**Structure**	205
	4.1 Wide-angle X-ray diffraction	205
	4.2 Small-angle X-ray diffraction	208
	4.3 Scanning electron microscopy	210
	4.4 Transmission electron microscopy	211
	4.5 Microstructure	215

5		**Fracture Mechanisms**	217
	5.1	Tensile failure	217
	5.2	Flexural failure	219
	5.3	Compressive strength	222
6		**Applications**	222
	6.1	Composite properties	223
	6.2	Aerospace uses	223
	6.3	Non-aerospace uses	225
	6.4	Future trends	225
		References	226

Chapter 7

Mechanical Properties of Cokes and Composites

		Summary	229
1		**Introduction**	230
2		**Nature of Cokes and Composites**	231
	2.1	Types of cokes	231
	2.2	Influence of production conditions on coke properties	232
	2.3	Types of composites	232
3		**Mechanical Properties**	235
	3.1	Deformation and elastic properties	235
	3.2	Failure	237
	3.3	Brittle fracture	238
	3.4	Fatigue	238
	3.5	Abrasion and wear	238
	3.6	Effects of temperature	239
	3.7	Effects of rate of loading	239
4		**Test Procedures**	240
5		**Composite Materials**	244

6	**Theoretical Considerations**		246
	6.1	The Griffith concept	246
	6.2	Interfacial aspects	248
	6.3	Statistical aspects	249
7	**Structural Considerations**		251
	7.1	Influence of porosity in cokes	251
	7.2	Carbon texture in cokes	253
	7.3	Microstructure and fracture in cokes and composites	254
	7.4	Fractography	255
8	**Conclusions**		256
	References		256

Chapter 8

The Nature of Coal Material

	Summary		259
1	**Introduction**		260
2	**Occurrence of Coal**		261
	2.1	Distribution	261
	2.2	Coal seam properties	262
3	**Bulk Properties and Chemical Structure**		263
	3.1	Chemistry	263
	3.2	Thermal and fluid properties	265
	3.3	Coking properties	266
	3.4	Chemical structure	268
4	**Coal Composition**		269
	4.1	The maceral concept	269
	4.2	Vitrinite macerals	271
	4.3	Liptinite macerals	273
	4.4	Inertinite macerals	275

5	**Coal Rank**		275
	5.1	Coalification	275
	5.2	Commercial classifications by rank	276
6	**Problems Peculiar to Coal as a Material**		278
	6.1	Sampling	278
	6.2	Mineral matter	278
	6.3	Weathering	282
	References		283

Chapter 9

Coal to Coke Transformation

	Summary		285
1	**Introduction**		286
	1.1	History of coke making	286
	1.2	The by-product battery	288
	1.3	By-products from coking	289
2	**Theories of Carbonization**		289
	2.1	Solvent extraction theory	289
	2.2	Transient fusion theory	289
	2.3	Precursors of the metaplast theory	290
	2.4	Metaplast theory	290
	2.5	Liquid crystal theory	291
	2.6	Liquefaction theory	291
3	**Origin of Coal**		292
4	**Coal Classification**		292
5	**Coking in a Single Coal Particle**		293
	5.1	Coke cenosphere	296
	5.2	Mesophase concept	297
	5.3	Role of plastic layer in coking	298

	5.4 Thermal transformation of coal macerals (Anthrathermotics)	298
6	**Rheological Properties of Coal**	300
7	**Coal Petrography**	300
8	**Coke Strength Prediction**	302
9	**Coke Petrography**	305
10	**Coke Pore and Wall Structure**	309
11	**Coal Blends for Coke Making and Blast Furnace Coke Properties**	311
	11.1 Coal blend	311
	11.2 Carbonization variables	312
	11.3 Coke strength determinations	313
	11.4 Coking pressure	313
12	**Coke in the Blast Furnace**	314
	12.1 Coke reactivity	315
	12.2 Petrographic strength predictions for foundry coke	316
	12.3 Formed coke	317
	12.4 Pre-heating	317
	12.5 Co-carbonization	318
	References	318

Chapter 1

Structure in Carbons and Carbon Forms

I.A.S. Edwards

Northern Carbon Research Laboratories, Dept. Of Chemistry, University of Newcastle upon Tyne, Newcastle upon Tyne, NE1 7RU, U.K.

Summary.

This Chapter introduces the basic structural features of carbon materials, briefly describes and gives definitions of the various carbon forms met with in carbon science and reviews the principal techniques by which the structure of solid carbons can be investigated.

Carbon is an element with a unique ability to bond with itself principally via sp^3 (diamond-like) and sp^2 (graphite-like) hybridization. The resultant structures have an immense variety of possibilities but for most of the materials dealt with in carbon science they can be considered as composed of mainly graphitic sub-units, with more or less structural order, linked together by less ordered regions.

Two extremes of structural organisation are distinguished as graphitizable and non-graphitizable carbons. The former (cokes) come from carbonaceous precursors which pass through a liquid phase on pyrolysis (_e.g_. pitches), while the latter (chars) are formed by those which do not fuse (_e.g_. wood).

By utilizing techniques such as X-ray diffraction, optical and electron microscopy, etc., the structural features of carbon materials can be elucidated at various levels of resolution. An estimate of the average crystallite size and perfection can be made from X-ray measurements. The degree of ordering at the micrometer level can be determined using optical microscopy while electron microscopy can be used to reveal the structure at the graphitic plane level.

STRUCTURE IN CARBON AND CARBON FORMS

Ian A. S. Edwards

Northern Carbon Research Laboratories, Department of Chemistry,
University of Newcastle upon Tyne, NE1 7RU, U.K.

1. INTRODUCTION - Setting the Scene.

This first Chapter gives a broad overview of the basic structures encountered in solid carbons, the carbon forms available and methods used to investigate their structures. Much of the material presented here is expanded upon in later Chapters.

1.1 The element carbon

The element carbon has an atomic weight of 12.011 and is element number 6. Three isotopes are known ^{12}C, ^{13}C, ^{14}C. The natural abundance of the stable isotopes (CRC, 1985) is:

$$^{12}C - 98.90\% \qquad ^{13}C - 1.10\%$$

The radioactive isotope ^{14}C, which is generated in the upper atmosphere by neutron bombardment of nitrogen ($^{14}N + n = {}^{14}C + {}^{1}H$) has a half-life of 5730 years. As well as being used for dating archaeological artifacts, ^{14}C is useful as a tracer in the study of organic reactions. ^{13}C with its magnetic moment (spin $^1/_2$) is ideal as a probe for nmr studies.

Because of its large abundance and combining power, ^{12}C is used as the reference definition for atomic mass, being defined as having the Relative Atomic Mass of 12 exactly. All other atomic and molecular masses are now based upon this definition (Cameron and Wichers, 1962).

For most carbon science requirements, the isotopic composition of the carbon is irrelevant as the properties are governed by the electronic configuration. With its central position in the first full row of the periodic table, carbon exhibits unique bonding possibilities. In essence, these are relatively simple without the complication of d-orbital involvement.

The electronic ground state 1s², 2s², 2p² is almost unknown because of the energetic advantage of involving all four outer orbital electrons in bonding between carbon atoms themselves, or with other atoms. In fact, carbon displays catenation (bonding to itself) to a unique degree. The resulting structures of chains, rings and networks are the carbons which are the subject of this book.

1.2 Bonding in carbon materials

The formation of σ- and π- bonds between carbon atoms and with other atoms (e.g. N, O, etc.) leads to the possibility of extensive and complex structures manifest in a whole branch of chemistry (Organic Chemistry) devoted to carbon compounds. The stability of carbon bonds and, in particular, the multiple bonding available through π- bonds is a principal feature of Carbon Science.

<u>Bond Energies</u> (kJ mol⁻¹) (Pauling, 1960)

C-C	348	C=C	699	C≡C	962
C-H	435	C-O	351	C=O	581
C-S	259	C-N	292	C-F	441
C-Cl	328	C-Br	276	C-I	240

There are two ways of treating the bonding in carbon compounds both of which give more or less the same results as far as descriptions of the structures are concerned. The first invokes the principle of hybridization where the orbitals in the outer shell are mixed together to form hybrids:-

One s- and three p- orbitals → Two sp- and two p- orbitals
→ Three sp²- and one p- orbitals
→ Four sp³- orbitals

The hybrid orbitals can then be assumed to link with compatible orbitals on other atoms to form σ- bonds while the p-orbitals are free to form π- bonds.

The second approach is to consider the formation of molecular orbitals directly from the atomic orbitals. This can be illustrated by the diagram shown in Figure 1.

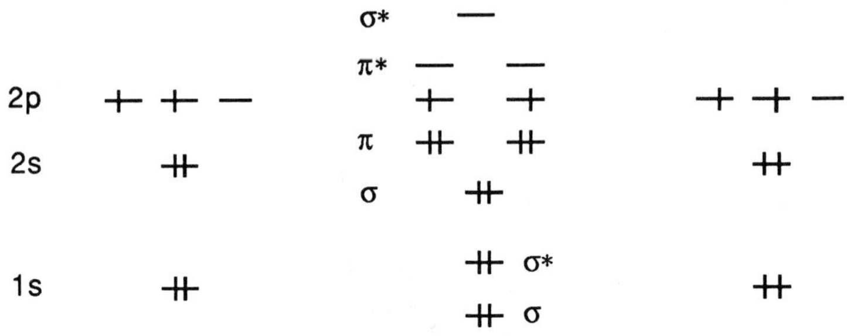

Figure 1. Molecular orbital diagram for C≡C bond.

Two principle bonding regimes dominate carbon compounds. For the materials of interest to the carbon scientist, the second of these predominates:-

1) σ-bond - diamond or aliphatic type, leads to chains of carbon atoms (e.g. paraffins) or three-dimensional structures which are quite rigid and isotropic.
2) σ-bond and π-bond - graphite or aromatic type, where the majority of structures are layered with a high degree of anisotropy.

Needless to say, most carbon materials contain both types and the degree of complexity is immense. A brief description of the two generic carbon forms is given first.

1.3 Diamond and graphite - perfect structures

The two regularly ordered allotropes of carbon are graphite and diamond. At ambient temperatures and pressures, graphite is the most thermodynamically stable:-

$$C(diamond) \rightarrow C(graphite) \quad \Delta H = -2.1 \text{ kJ mol}^{-1}$$

However, from a kinetic point of view the change is extremely slow at room temperature (rapid at about 1900 K) because of the large number of bonds which would need to be broken in the process.

Diamond is made up of a regular three-dimensional network of σ-bonds providing a very rigid, stable structure (Figure 2) making it the hardest material known. Thermodynamically, it is more stable than graphite at pressures > 60 GPa at room temperature principally due to its higher density (3.51 g cm^{-3}) compared with that of graphite (2.25 g cm^{-3}) (CRC, 1985). Within the diamond lattice, the bonding electrons are fixed between atoms so that electrical conductivity is very low.

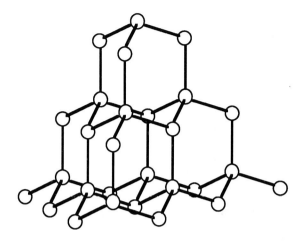

Figure 2. Structure of diamond.

In graphite, there is both σ- and π- bonding holding the atoms in hexagonal two-dimensional networks. These layers are held rather loosely together by van der Waals forces. The stacking of the layers is principally - A.B.A.B., the hexagonal form, illustrated in Figure 3, with a small proportion (< 10%) in natural graphites stacked - A.B.C.A.B.C., *i.e.* the rhombohedral form.

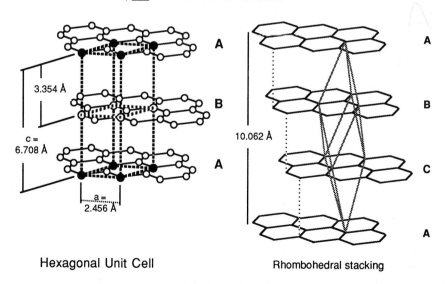

Hexagonal Unit Cell   Rhombohedral stacking

Figure 3. Graphite structure.

The energy needed to slide the layers over one another is low making graphite a soft material which can be used as a lubricant. This feature also means that actions such as grinding can increase the amount of rhombohedral form present in the graphite. The rhombohedral form irreversibly converts to the hexagonal form at ~ 2400K and, therefore, is not present in synthetic graphite (Boehm and Coughlin, 1964).

The conjugated π- bonding of the layered structure results in the electrons being delocalised throughout the structure providing a means of conducting electricity similar to metallic conduction bands. Across the layers, however, there is no electron movement so conduction in this direction is a minimum. This is an excellent example of the anisotropy exhibited by graphite.

2. ORDER / DISORDER

Carbon materials are, in general, a mixture of well-ordered material, often of short range (<100 nm), surrounded by less-ordered material. The proportions of and relationship between the ordered and less-ordered parts is the main topic of this Chapter.

2.1 More-ordered structures

The principal parts of the more-ordered structures are essentially graphitic. Small volumes often have an almost perfect graphite lattice but, as the volume increases, the presence of defects, distortions and heteroatoms destroys the regularity giving material which is very disordered. True single crystal graphite is vary rare. Natural flake graphite, _e.g._ Tyconderoga, has high crystalline order but probably the most perfect graphites are the highly ordered synthetic pyrolytic graphites (Moore, 1981).

All graphites contain defects within their structures, _i.e._ stacking faults, dislocations, vacancies and interstitial atoms (Amelinckx et al., 1965). Stacking faults, leading to a ribbon of rhombohedral graphite within the overall hexagonal stacking, require no C-C bond breakage. However, dislocations, of both line and screw types, necessitate the disruption of the lamellae. These, together with point defects, vacancies and impurities, explain the non-planarity of most of the carbon layers which make up carbon materials and are responsible for the cross-linking or pinning between the layers.

The study of defects is important for the understanding of mechanisms of gasification (Baker, 1982). Mild oxidation can be used to reveal the presence of some defects in carbon structures on examination by optical microscopy.

As mentioned earlier, little energy is needed to slide graphitic layers over one another. Similarly, twisting them so that they are not aligned with one another is also possible leading to structures which have roughly parallel and equidistant layers but with random orientation. These are sometimes called "turbostratic" carbons. The spacing between such layers is higher than that for oriented layers and similar to that brought about by other defects.

The degree of ordering within graphitic material has been the subject of extensive study. The pioneering work of Franklin (1951a) has had some later modification but her equation relating the d-spacing between the layers, measured by X-ray diffraction, to the degree of randomness in the alignment of the layers, p, is still valid:

$$d_{(002)} = 3.440 - 0.086(1-p^2)$$

The models which she proposed for carbon structure (Figure 4) have also stood the test of time and are close to the more elaborate descriptions which have since been attempted (cf. Oberlin, 1984).

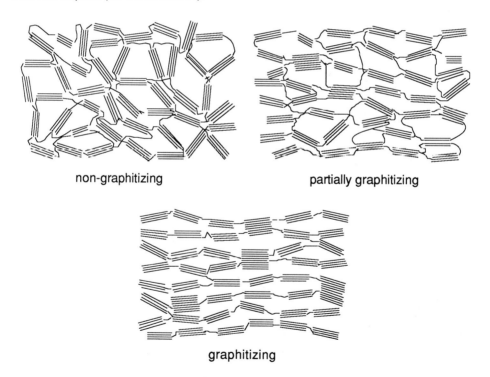

Figure 4. Franklin (1951b) models of carbon structures.

The order in sp³-based parts of most carbon materials is of even shorter range, although much of the less-ordered parts of the structure are aliphatic in nature. Again, defects, distortions and heteroatoms are present within any extended structure.

2.2 Less-ordered structures

Most carbon materials are composites in that they are made up of entities which can be separately identified within the overall structure. The principal constituents can be said to be:

a) ordered or graphite-like where there is a high degree of anisotropy.
b) less-ordered where for most tests the structure is isotropic.

The processes of conversion of less-ordered into more-ordered material and vice versa are the principal concerns of the carbon industry, whether by heat treatment, ageing or other energy input. These processes will be considered elsewhere in the book. However, the progressive increase in order during carbonization of a material, such as pitch coke, resulting in a highly graphitic carbon serves to illustrate the range of structures encountered in carbon science (Marsh and Griffiths, 1982). (Figure 5)

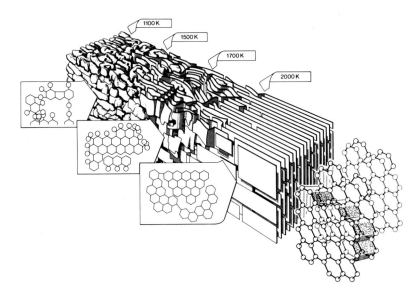

Figure 5. Marsh-Griffiths model of carbonisation/graphitisation process.

2.3 Range of order

The ordered material within carbons can be thought of as crystallites of graphite and, by measuring the size of these crystallites, one estimate of the range of order can be made. However, the orientation of crystallites may be such that a more appropriate estimate of the range of order may be many orders of magnitude higher than the size of the crystallites themselves.

The kind and range of ordering considered within the structures may depend upon the investigating technique used. In a later section, methods for studying carbon structures will be reviewed.

The degree of isotropy and anisotropy present in carbon materials is of importance when attempting to assess their properties and performance. For example, macroscopic properties of many carbons are isotropic while, on a microscopic level, the structure presents a high degree of anisotropy. Also the individual parts of the structure may be anisotropic to one examination technique but isotropic to another. An example of this is the optical texture of a carbon when compared with its electron microscope image. Using optical microscopes, features with dimensions less than ~ 0.5 µm cannot be resolved and, therefore, some parts are described as isotropic when the electron microscope shows these to be highly anisotropic at the nanometer level.

Heteroatoms within the carbon skeleton are another manifestation of disorder, in this case, at the molecular level. Hydrogen, oxygen, nitrogen, sulphur and halogens are present to some extent in nearly all natural sources of carbon materials. The presence of other material, such as mineral inclusions or metallic particles, add to the amount of disorder in carbons. Another aspect which is a consequence of the layered structure of graphite is that atoms, groups and molecules can be intercalated between the carbon layers (Moore, 1981). In some cases, stoichiometric compounds are formed with graphite (Figure 6) while, in very disordered carbons with distorted lamellae, intercalation can be considered as a special case of adsorption in ultra-microporosity.

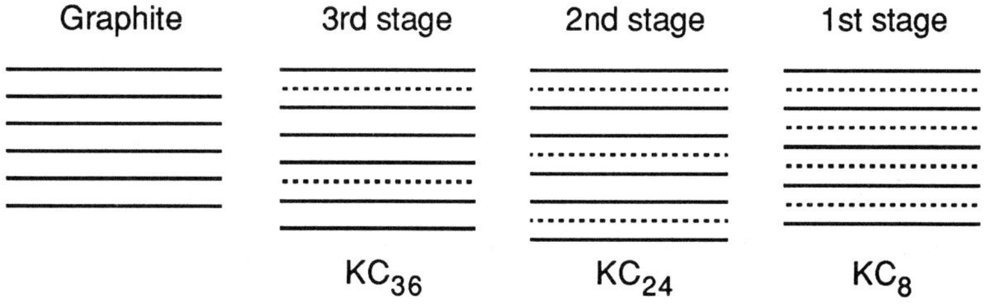

Figure 6. Intercalation of potassium in graphite.

3. CARBON FORMS

Carbon Science deals with solid carbons almost all of which (with the notable exception of diamond) have the basic structural unit of planar or approximately planar arrangements of carbon atoms in a hexagonal network. The solid carbons essentially exist as Graphitic and Non-Graphitic forms.

Definitions given below are based on the "tentative definitions" suggested by the International Committee for Characterization and Terminology of Carbon (ICCTC). These are published periodically in the Journal, Carbon; six lists have been produced (ICCTC, 1982, 1983, 1985, 1986, 1987a & 1987b).

Solid carbons are mainly derived from organic precursors by pyrolysis, the process being termed Carbonization.

DEFINITION **Carbonization** is a process of formation of material with increasing carbon content from organic material, usually by pyrolysis, ending with an almost pure carbon residue at temperatures up to 1600 K (ICCTC, 1982).

3.1 Graphitic carbons (natural and synthetic graphites)

DEFINITION **Graphitic carbons** are all varieties of substance consisting of the element Carbon in the allotropic form of Graphite irrespective of the presence of structural defects (ICCTC, 1982).

DEFINITION **Graphite** is an allotropic form of the element Carbon consisting of layers of hexagonally arranged carbon atoms in a planar condensed ring system. The layers are stacked parallel to each other. The chemical bonds within the layers are covalent with sp^2 hybridization. The bonds between the layers are of van der Waals type (ICCTC, 1982).

DEFINITION **Natural graphite** is a mineral consisting of Graphitic Carbon regardless of its crystalline perfection. Some show a high degree of perfection, e.g. Tyconderoga graphite, but most commercial natural graphites mined are flake graphites containing other minerals (ICCTC,1982).

DEFINITION **Synthetic graphite** is a material consisting mainly of Graphitic Carbon which has been obtained by means of Graphitization Heat Treatment of Non-Graphitic Carbon or of chemical vapour deposition (CVD) from hydrocarbons at temperatures above 2100 K (ICCTC, 1982).

3.2 Non-graphitic carbons and graphitization.

DEFINITION **Non-graphitic carbons** are all varieties of substances consisting mainly of the element Carbon with two-dimensional long range order of the carbon atoms in planar hexagonal networks, but without any measurable crystallographic order in the third direction (c-direction) apart from more or less parallel stacking (ICCTC, 1982).

Many non-graphitic carbons can be converted into graphitic carbons by heat treatment to above 2500 K. Such conversion is called graphitization.

DEFINITION **Graphitization** is a solid state transformation of thermodynamically unstable Non-Graphitic Carbon into Graphite by thermal activation (ICCTC, 1982).

The degree of graphitization depends upon the temperature of the heat treatment and the time allowed to anneal structure.

3.3 Graphitizable and non-graphitizable carbons.

DEFINITIONS **Non-graphitizable carbons** are those which cannot be transformed into Graphitic Carbon solely by heat treatment up to 3300 K under atmospheric or lower pressure. **Graphitizable carbons** are those which are so converted (ICCTC,1982).

Non-graphitizable carbons are produced from wood, nut-shells and non-fusing coals. Essentially, the macromolecular (polymeric) structure of these materials remains during heat treatment only losing small molecules by degradation and developing even more cross-linking so that fusion cannot take place. The loss of small molecules and the retention of the complex macromolecular structure leads to high microporosity, with surface areas in the order of 1000 $m^2 g^{-1}$.

Most graphitizable carbons pass through a fluid stage during carbonization. This allows large aromatic molecules to align with each other so forming the mesophase precursor of the graphitic structure which is essential to the development of the coke (Marsh and Walker, 1979).

3.4 Pitches

Pitches are carbonaceous materials derived from organic precursors by relatively low temperature processes, _e.g._ distillation at < 700 K. They contain a large range of molecular types and masses. As the precursors to most carbon artifacts, their behaviour on carbonization has been the subject of much study and will be dealt with in later Chapters. Most pitches melt on heating to give an isotropic fluid. As heating is continued above 660 K, alignment of lamellar molecules occurs leading to the formation of nematic discotic liquid crystals. The further development of this liquid crystal system (<u>MESOPHASE</u>) provides the basic micro-structure of the final carbon material, controlling its optical texture (Brooks and Taylor, 1969, Honda, 1988, Lewis and Lewis, 1988 and Rand, 1985).

During the process of carbonization, volatile matter is released from the bulk of the material. This, together with the complex packing within and between the carbonized particles, results in a final material which is porous. The degree and nature of the porosity depends upon the precursor and the conditions of the carbonization process. These and other aspects of porosity will be dealt with in later Chapters.

Two principal types of pitch are used for the preparation of the bulk of carbon materials:

DEFINITION   **Coal-tar pitch** is a residue produced by distillation or heat treatment of coal tar. It consists of a complex mixture of numerous predominantly aromatic hydrocarbons and heterocycles (ICCTC, 1985).

DEFINITION   **Petroleum pitch** is a residue from heat-treatment and distillation of petroleum fractions. It consists of a complex mixture of numerous predominantly aromatic and alkyl substituted aromatic hydrocarbons (ICCTC, 1985).

Cokes from petroleum are, in general, more highly graphitizable than those from most commercial coal-tar pitches, the latter having a more subdivided mesophase structure caused by a larger concentration of insolubles (called QI material) in the parent pitch. Removal of the insolubles considerably enhances graphitizability.

3.5 **Cokes**

DEFINITION   **Coke** is a highly carbonaceous product of pyrolysis of organic material at least parts of which have passed through a liquid or liquid-crystalline state during the carbonization process and which consists of non-graphitic carbon (ICCTC, 1982).

Most coke materials are graphitizable carbons. Their structure is a mixture of various sizes of optical texture, from the optically isotropic to domain and flow anisotropy (~200 μm diameter). Only the short range order, associated with non-graphitic carbons, usually exists at the crystallographic level.

Various types of coke can be DEFINED.

Green coke is the primary solid carbonization product obtained from high boiling carbon fractions at temperatures below 900 K (ICCTC, 1983). (Also termed raw coke).

Calcined coke is a Petroleum or Coal Derived Pitch Coke with a mass fraction of hydrogen less than 0.1wt.%. It is obtained by heat treatment of Green Coke to about 1600 K (ICCTC, 1983).

Petroleum coke is a carbonization product of high boiling carbon fractions obtained in petroleum processing (ICCTC, 1983).

Coal derived pitch coke is the primary industrial solid carbonization product from coal-tar pitch (ICCTC, 1983). (This is often termed coal-tar pitch coke).

Metallurgical coke is produced by carbonization of coals or coal blends at temperatures of up to 1400 K to produce a macroporous carbon material of high strength (ICCTC, 1983).

Delayed coke is a commonly used term for a primary carbonization product of high boiling hydrocarbon fractions by the delayed coking process (ICCTC, 1983). Delayed coke has a better graphitizability than cokes produced by other coking processes even if the same feedstock is used. The principle products of delayed coking are sponge coke and needle coke. Shot coke is also produced as `agglomerates of spheres 1-2 mm diameter, but has no commercial value.

Sponge coke has a non-orientated anisotropic optical texture and is used as a filler for electrodes in the aluminium industry.

Needle coke is a commonly used term for a special type of coke with extremely high graphitizability resulting from a strong preferred orientation of its microcrystalline structure (ICCTC, 1983). This ordered anisotropy, indicative of its low coefficient of thermal expansion, is utilised in electro-graphites.

3.6 Coals

Coals possess a wide range of structures at both the microscopic and the molecular levels. They result from Coalification (Hirsch, 1954) of organic materials, mainly of plant origin. This is a geological and chemical process of dehydrogenation, deoxygenation and condensation which occurs in the earth's crust by gradual transformation at moderate temperatures (~ 500 K) and high pressures (Stach, 1982).

The development of coalification with time leads to a definition of the degree of coalification as the C/H ratio. From peat through lignites, sub-bituminous and bituminous coals to semi-anthracites and anthracites, the carbon content increases from 50 wt.% to over 95 wt.%. Because of the variations in conditions and original plant forms, the degree of coalification and the properties of coal show large variations with locality throughout the world.

At the molecular level, attempts have been made to describe the structure of coal incorporating the organic entities and functional groups in proportion to their determined presence (Shinn, 1984 and Gundermann, 1987). This has been partially successful for narrow ranges of coals but the complexity of structural types precludes a single widely applicable model. The interpretation of coal structure as a macromolecular network incorporating small, entrapped molecules is, perhaps, a better approach (Given 1988).

Coal petrography investigates coal structures at the microscopic level. By this means, structures which are based on the original plant material can be identified, together with non-carbon materials (Stach, 1982). The latter are mostly minerals or of mineral origin. The carbon entities are termed macerals.

Both the nature of coal and the conversion of coal to coke are dealt with in detail in Chapters 8 and 9. However, a major use of coal, namely combustion, leads to conflict in the use of the term CHAR. In studying coal combustion, a char is the product of rapid pyrolysis, essentially in an inert or reducing atmosphere, and is considered to be the intermediate after the first stage of the combustion process. In this case, some of the material may have passed through a fluid state. In contrast, the recommended definition for carbon science is:

DEFINITION   **Char** is a carbonization product of a natural or synthetic organic material, which has not passed through a fluid stage during carbonization (ICCTC, 1983).

It is important when using this term that this distinction should be clear. In both cases, however, the char form often depends upon the shape of the precursor, leading to a highly porous solid material.

3.7 Carbon fibres

DEFINITION   **Carbon fibres** are fibres (filaments, yarns, rowings) consisting of at least 92% (mass fraction) carbon, regularly in non-graphitic stage (ICCTC,

1987a). The term "Graphitic Fibres" is justified only if 3-dimensional crystallographic order is confirmed by X-ray diffraction measurements.

The anisotropic properties of carbon materials are exploited by the alignment of the lamellar planes along the fibre axis but without producing long-range order normal to the planes.

There are two principal types of carbon fibre; those based on PolyAcryloNitrile (PAN) (Bennett et al., 1983, and Johnson, 1987) and those based on mesophase pitch (Donnet and Bansal, 1984). Both are considered in detail in Chapter 6.

3.8 Other carbon materials

DEFINITION   **Charcoal** is a traditional term for a char obtained from wood and some related natural organic materials (ICCTC, 1983). It retains the form of the parent material, having in many cases a highly developed pore structure.

The basic structure is disordered at the microscopic level resulting in isotropic optical texture. At the crystallographic level, the order is still low with no detectable graphitic properties.

DEFINITION   **Carbon blacks** are industrially manufactured Colloidal Carbon materials in the form of spheres and of their fused aggregates with sizes between 10 and 1000 nm (ICCTC, 1985). The structural order found in carbon blacks varies with the method of preparation but, in general, there is alignment of the carbon layers parallel to the surface of the spheres (Donnet and Voet, 1976).

Deposited carbons are produced by chemical vapour deposition (CVD) of carbon from volatile hydrocarbon compounds on carbon, metal or ceramic substrates. This provides a method for obtaining carbon materials with a homogeneous microstructure.

In cases where the substrate is an active catalyst for carbon deposition, the growth of whiskers or filaments by a solution/deposition mechanism, with small metal particles at the tip, has been observed (Baker and Harris, 1978 and Bradley et al., 1985). These filaments are often highly graphitic.

DEFINITION   **Activated carbons** are porous carbon materials, usually chars, which have been subjected to reaction with gases during or after carbonization in order to increase porosity (ICCTC, 1983).

The degree and type of porosity can be controlled to provide materials with large and, in some cases, specific adsorption capacity. Carbon materials with a high surface area are also ideal as substrates for the support of catalysts.

Mesocarbon microbeads is the term introduced by Honda and Yamada (1973) to describe the mesophase spheres generated on heat-treating pitches and separated by solvent extraction or other means. Work by Auguie et al. (1980) has shown them to have the classical Brooks and Taylor structure. The control of their size and morphology is an expanding field of study (Mochida et al. 1985 and Shimokawa et al. 1986). Mesocarbon microbeads have many potential applications in mechanical carbons, as filters and as adsorbates (Kodama et al., 1988).

Diamond-like films, the development of which has taken place over the past few years, have provided much interest in structures which have little graphitic character. These films are usually produced by arc discharges in hydrocarbon gases in the presence of hydrogen. The resultant carbon film deposited on a suitable substrate has diamond-like structure. The films exhibit the properties of diamond, e.g. hardness, and can therefore be used for abrasion resistant coatings (Messier et al., 1987)

4. **COMPOSITES**

All above carbon materials are very heterogeneous but, being from individual sources, are homogeneous in terms of treatment. True composites can be defined as artifacts where two or more forms have been brought together to produce a material nominally, at least, of more than one component.

The ease with which the components of a composite can be identified varies with the nature of the original carbon form and the subsequent processing of the material. In some cases, simple optical microscopic examination is sufficient, identifying differences in the optical texture or morphology of the components, e.g. in green anodes, the binder-pitch semi-coke can be distinguished from the grist or filler coke.

For more intimately mixed materials and those where subsequent heat treatment has produced a relatively uniform optical texture, etching of the polished surface followed by SEM examination can reveal the differences in reactivity of the components. Etching can be accomplished by exposure to acid, e.g. chromic acid, partial gasification using carbon dioxide or attack by atomic oxygen (Marsh et al. 1981).

Where the intermingling of the components is below the resolution of the optical microscope or standard SEM instruments, selected area electron diffraction can give some indication of the presence and position of the more ordered crystalline components.

In cases where a high degree of graphitization has taken place in the production of the composite, it can be impossible to identify the structures resulting from the individual components.

NOTE All these techniques are also applicable to the investigation of ordinary "single" component carbon materials and can reveal the diversity of structures present in them.

One aspect of composite materials which is of great interest is the interface between components, e.g. in the carbon electrode to be considered below. The strength of the composite, is often governed by the adhesion forces which can be both chemical or physical in nature.

4.1 Graphitic composites

These materials are prepared from a mix of a highly graphitizable filler coke, such as needle coke, and a binder which is usually a coal-tar pitch. The mix is, usually, pre-baked to provide a cohesive shape which can then be graphitized to temperatures up to 3000 K. In some cases, the prebaked (or green) intermediate is densified by impregnating, under pressure, with extra binder. For large artifacts, graphitization is accomplished by surrounding the piece with coke to provide a relatively reducing environment and heating electrically. Smaller items, particularly when high purity is needed, are carbonized in electrically powered carbon resistance furnaces which are purged with inert gas (Mantell, 1968).

Graphite electrodes are massive artifacts, mainly used in the production of steel. They are highly graphitic, exhibiting an anisotropic optical texture, and capable of carrying a large electrical current at high temperature and under large thermal stress.

For nuclear graphite the criteria is for high chemical purity, to avoid adsorption of low-energy neutrons, and no preferred bulk orientation of the graphitic crystallites, conferring high dimensional stability. The resulting optical texture is, therefore, mostly fine grain with random orientation.

4.2 Carbon electrodes

The choice of components for carbon electrodes which do not need to be graphitized is controlled mainly by cost and availability. For electrodes used in aluminium smelting, the filler is calcined, delayed petroleum coke and the binder a coal-tar pitch.

The processing is similar to that for graphitic electrodes, with mixing, shaping, prebaking, densifying (in some cases) and heat treatment. The heat treatment temperature, however, is much lower, usually below 2000 K.

Carbon electrodes in aluminium production need to fulfil two tasks; to provide the means for electrical transport to the cell and to act as the reductant in the electrochemical process, i.e. conversion of alumina to aluminium. They must also be resistant to gaseous oxidation to reduce losses by the Boudouard reaction. The role of the binder pitch, therefore, is to provide, on carbonization, coke bridges between the pieces of filler coke which will hold the structure together and give electrical contact.

To achieve this, the pitch must be fluid enough to enter the pores of the filler during baking but not so fluid that it flows away from around the particles. Good anodes exhibit successful binding which is obtained by the coke bridges keying into the surface of the filler providing a good mechanical bond which allows the passage of electricity (Latham and Marsh, 1986).

4.3 Carbon/carbon composites

While all the composites already considered are carbon/carbon composites, the term usually refers to either carbon fibre reinforced carbon artifacts or high performance materials. These high performance carbon materials are usually made by subjecting the mixed components to high pressure during carbonization.

Carbon fibres held in a pitch matrix give a high strength material with excellent thermal and mechanical shock resistance, the key to the strength being the adhesion of the pitch coke to the carbon fibres. During carbonization, the pitch mesophase aligns with the fibre surface. Therefore, it is important that the thermal properties of the pitch coke are as close to those of the fibres as possible. Failure to match them results in cracks and fissures at the interface so weakening the bulk structure. Carbon fibre reinforced polymer composites are a major use of carbon fibres but cannot strictly be called carbon/carbon composites.

Isostatic pressing of carbon materials during carbonization results in artifacts of high density and no bulk preferred orientation. By control of the conditions, suitable fine grain optical texture can be obtained, conferring high strength.

5. METHODS OF STUDYING CARBON STRUCTURE

5.1 Optical microscopy

Specimens are prepared by mounting the sample in a resin block and polishing the surface to optical flatness using alumina or diamond paste (if the specimen is massive enough it can be polished without mounting). The polished surface is examined using reflected light microscopy usually with polarized light. To observe interference colours, due to the orientation of the graphitic lamellae at the surface, parallel polars are used with a half-wave retarder plate between the specimen and the analyser (Forrest and Marsh, 1977). The general arrangement is shown in Figure 7 together with a diagrams illustrating the generation of the interference colours.

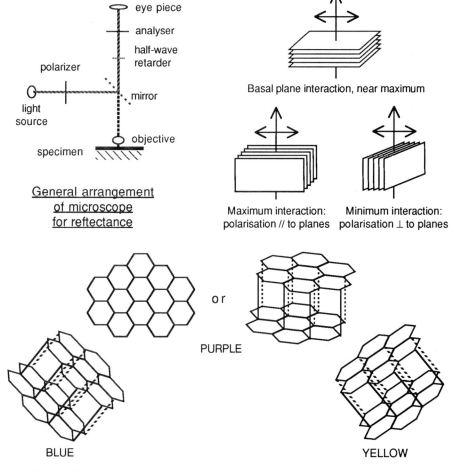

Figure 7. Polarised light optical microscopy and interference colours

The appearance of a surface is called its OPTICAL TEXTURE. Figures 8, 9 and 10 illustrate the type of textures observed. The size and shape of the isochromatic areas can be estimated. Dimensions vary from the limit of resolution ~0.5 μm to hundreds of micrometers. The nomenclature used to describe the features has been developed over many years (Patrick, et al. 1973, Grint, et al. 1979 and Moreland, et al., 1988) and discussions are underway to establish a more standard system. The Table below gives the latest definitions of the classes of optical anisotropy together with a definition of Optical Texture Index (OTI) factor which is useful for characterizing a carbon material.

Nomenclature used to describe Optical Texture		OTI factor
Isotropic (Is and Ip)	No optical activity	0
Fine mosaics (F)	< 0.8 μm in diameter	1
Medium mosaics (M)	> 0.8 < 2.0 μm in diameter	3
Coarse mosaics (C)	> 2.0 < 10.0 μm in diameter	7
Granular flow (GF)	> 2 μm in length; > 1 μm in width	7
Coarse flow (CF)	> 10 μm in length; > 2 μm in width	20
Lamellar (L)	> 20 μm in length; > 10 μm in width	30

By point-counting the individual components of the microscope image and multiplying the fraction of points counted for each component by the corresponding OTI factor and summing the values, the optical texture index for the sample can be obtained. This number gives a measure of the overall anisotropy of the carbon.

It must be stressed that this is only a comparative technique and that material characterized as isotropic may only be so at the level of resolution. This becomes obvious when a surface is examined with a higher grade objective (not just higher power) allowing more anisotropy to be observed (distinguished).

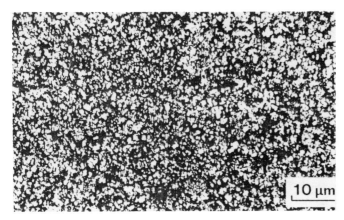

Figure 8   Optical micrograph of coke surface - fine grained mosaic, OTI = 1

Figure 9   Optical micrograph of coke - medium and coarse mosaic, OTI = 3

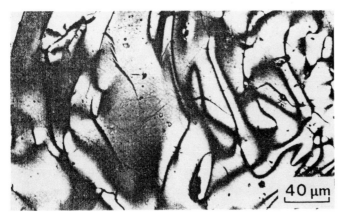

Figure 10   Optical micrograph of coke surface - coarse flow, OTI = 30

Other uses of optical microscopy in the study of carbon materials are principally concerned with coal petrography, a discipline based on the measurement of reflectance from the coal surface and dealing with coal materials in terms of macerals. These are forms identifiable as deriving from the original plant material via the coalification process (Stach, 1982) illustrations are given in Chapter 8 indicating the range of forms found in coals. An area related to this is the study of char forms derived from the pyrolysis of coal. The nomenclature used to describe such materials is at present being discussed by the ICCP (International Conference on Coal Petrology)

5.2 Electron microscopy (SEM and TEM)

Both the two major types of electron microscopical examination of carbons present excellent opportunities for structural study and, equally, problems in specimen preparation and interpretation of the images produced.

Scanning Electron Microscopy (SEM) uses the scattering of electrons from the surface of the sample to reveal the topography of material.(see Figure 11). It also utilises the scattering power of different materials to distinguish them.

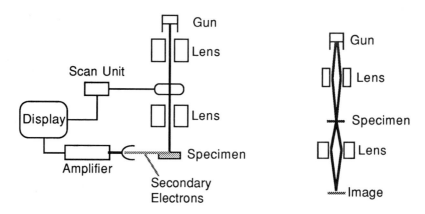

Figure 11   Scanning electron microscopy and transmission electron microscopy.

Gold coating the specimens gives a much better resolution of the topographical features. Unfortunately, this procedure cannot be used in conjunction with another useful SEM facility which uses the interaction of the electron beam with some of the surface atoms (up to 100 nm deep) to give X-rays which can be analysed to reveal the surface composition. One such technique uses energy dispersal to analyse the X-rays emitted and is usually called EDAX. This technique is limited in most general instruments to elements of higher atomic weight than fluorine.

Carbon surfaces which have been polished for optical microscopy show very few features in a SEM examination (no topography). However, etching the surface either chemically (with chromic acid) or by ion bombardment reveals a wealth of detail which can be related to the optical texture of the sample. A specialised application of this is the "same area" technique where a specific part of a polished surface which has been identified and characterized optically is re-examined by SEM following etching (Markovic and Marsh, 1983). Figures 12 and 13 show micrographs illustrating this technique.

SEM is an excellent method for monitoring the changes in topography following various treatments, such as - gasification, heat treatment, etc. Figures 14 and 15 are micrographs of metallurgical (blast furnace) coke before and after reaction with carbon dioxide.

Transmission Electron Microscopy (TEM) provides a means of obtaining high resolution images of carbon material. Figure 11 includes an outline of the general arrangement of a transmission electron microscope.

The preparation techniques for TEM are quite difficult. It is necessary to obtain a very thin section of the carbon, less than 100 nm, of a uniform thickness. Specimen breakage can often give good results but this should be treated with caution as it can lead to random variations in thickness which cannot be fully interpreted in terms of the image produced. A more controlled method of producing suitable material is by cutting a thin section with a microtome and further thinning the centre portion by ion bombardment. The uniform thickness is important because, in these investigations, it is the variation in the amount of material through which the electron beam passes that provides the image contrast and fringe imaging can be an artifact of a tapered sample. When the conditions are correctly established, high resolution TEM can provide direct imaging of the layer planes in carbon materials, reveals the complexity of the most regular structures and shows the ordering present down to the nanometer level (Marsh and Crawford, 1984 and Bourrat, et al., 1986). cf. Chapter 6.

Figures 16 and 17 are micrographs showing the resolution of layer planes in both highly ordered and less ordered material obtained by high resolution transmission electron microscope (HRTEM).

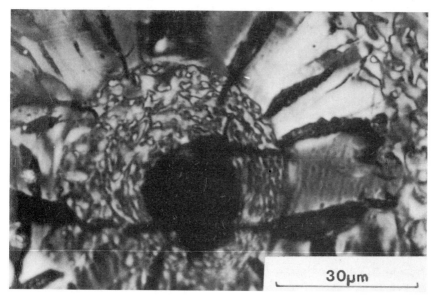

Figure 12   Optical micrograph of calcined shot coke - before etching.

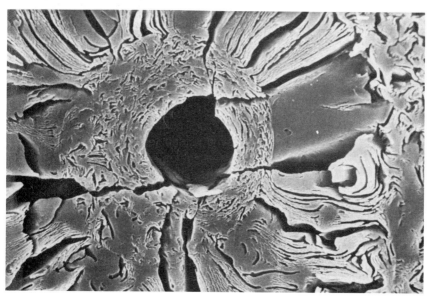

Figure 13 SEM micrograph of calcined shot coke - after chromic acid etching.

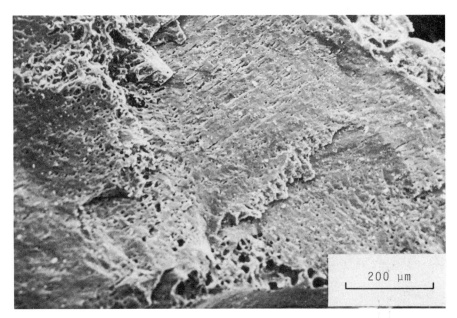

Figure 14   SEM micrograph of blast furnace coke - original surface.

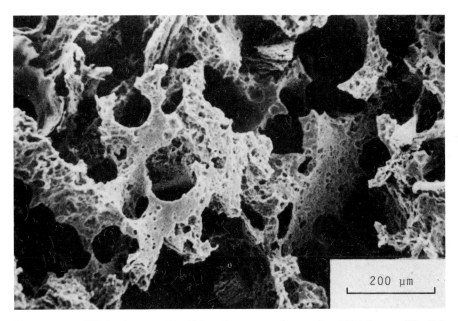

Figure 15   SEM micrograph of blast furnace coke - after 75% burn off in CO_2.

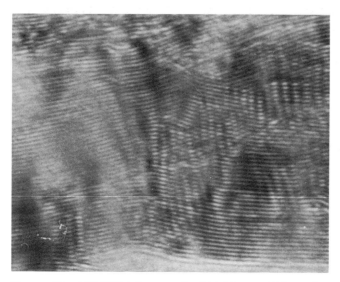

Figure 16   HRTEM micrograph of highly graphitic carbon - highly ordered.

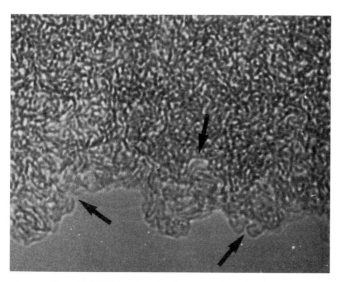

Figure 17   HRTEM micrograph of PVDC carbon (HTT 1473 K) - disordered.

5.3 X-ray diffraction

The average bulk structure of carbon materials can be readily revealed using X-ray diffraction. Early examples of such work on coals, chars and cokes are given by Blayden, et al. (1944) and extended to graphites in the review by Riley (1947). More recently Kawamura and Bragg (1986) have combined X-ray diffraction with strain and weight changes to follow the graphitization of pitch coke. The technique provides a measure of the amount of ordered material present and can be used to give an indication of the size of the crystallites which make up the ordered structure.

The samples are usually prepared as powders either in capillaries or spread on a flat sample holder. The X-ray diffraction pattern is recorded either on film or with a diffractometer. Figure 18 shows the arrangement of a powder diffractometer.

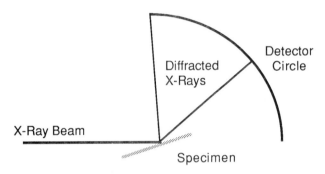

Figure 18. X-ray diffraction.

The resultant pattern is the amount of scattering over a range of scattering angles, θ, and can be analysed in terms of diffraction peaks, their positions and their widths. For the most accurate work, a standard (usually a highly crystalline salt) is added to the powder to provide internal calibration of the peak positions and widths so permitting any instrumental factors to be taken into account.

Although the principal scattering derives from the ordered material present, some indication of the amount of disorder can be obtained by the background scatter. Similarly, the broadening of the diffraction peaks allows an estimation of the mean particle size to be made (Jeffrey, 1971). Line-broadening arises from both the strain or defects in the lattice and the finite crystal size. Assuming that the defects in the lattice reduce the extent of order, an effective crystallite size, t, can be estimated from the amount of broadening, β, using the Sherrer equation:

$$t = \kappa\lambda / \beta \cos\theta$$

where λ is the wavelength θ is the scattering angle.

and the value of κ (~ 1) depends opon the shape of the crystallite.
<u>e.g.</u> κ has the values 0.9 for L_c and 1.84 for L_a.

β is the amount of broadening due to the sample and the observed broadening, B, usually needs to be corrected for the instrumental broadening, b, using relationships such as:

$$\beta^2 = B^2 - b^2$$

The measurement of line broadening is illustrated in Figure 19.

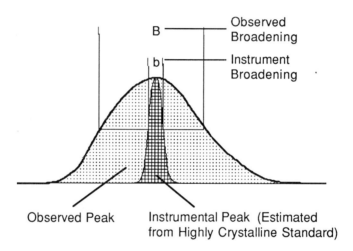

Figure 19   X-ray diffraction line-broadening.

The parameters usually quoted from X-ray diffraction experiments are:

   $d_{(002)}$ interlayer spacing. L_c stack height. L_a stack width.

These give some indication of the degree of graphitization of the carbon.

One specialized area of diffraction, which will be expanded upon later in the book is that of fibre diffraction (see Chapter 6). Both normal and small angle scattering patterns are used for this investigation. Carbons with large surface area can also be successfully investigated by a combination of small (low) and normal (high) angle scattering (Konnert and D'Antonio, 1983). Small angle scattering has also been used to give some indication of pore structure (Dubinin and Plavnik, 1968 and Janosi and Stoeckli, 1979).

5.4 Raman spectroscopy

The vibrational characteristics of carbonaceous material exhibited in Raman spectra allow distinct resonances from ordered and disordered regions of the material to be identified (Mernagh et al., 1984, Lespade et al., 1984 and Cottinet et al., 1988). Ordered graphitic layers show resonance between 1580 cm^{-1} and 1600 cm^{-1}. This band narrows and moves closer to 1580 cm^{-1} (the resonance for graphite) with graphitisation. A second resonance, 1350 - 1380 cm^{-1}, disappears with graphitisation and is assigned as a defect band.

5.5 Surface techniques

A number of surface sensitive techniques have been applied to carbon materials. However, the heterogeneity of most surfaces makes a full analysis of such surfaces as difficult as that of the bulk carbon structures.

Low Energy Electron Diffraction (LEED) can be applied with any degree of understanding only to regular crystal surfaces (Clarke, 1985). LEED patterns of single crystal graphites have been obtained and some attempt made to study changes in these surfaces following treatment such as gasification.

Auger Electron Spectroscopy (AES), which can often be used in conjunction with LEED, has been of value as it can be used to study carbon deposition on both single crystal and polycrystalline metal surfaces (Komiyama, et al., 1985).

Probably the most useful of these techniques is X-ray Photoelectron Spectroscopy (XPS) (Kozlowski and Sherwood, 1986) - sometimes also called ESCA (Electron Spectroscopy for Chemical Analysis). With this method, the electronic state of atoms can be probed so that the bonding configuration in the case of surface species can be determined. This, for instance, can reveal the difference between sulphur bound to the surface in sulphide or in sulphate groups and the presence of carbonyl or carboxyl groups (Brown, et al., 1981).

All these techniques have one major drawback; they can only be applied under high vacuum conditions. This means that in situ changes in the surface due to oxidation, etc. cannot be observed directly.

6. FACTORS IN CARBON STRUCTURES

All carbon forms, apart from diamond, have graphitic lamellae. They differ from each other in the size of these lamallae, the perfection and planarity of the layers and their relative stacking, the relative amount of disordered material and porosity which is present. An estimate of the degree of order present can be found from X-ray diffraction measurements but these provide only an average for the whole material. A more detailed examination using high resolution electron microscopy reveals the microstructure over a small sample of the material. To obtain as detailed a description as possible, the application of a range of investigations is desirable, this being exemplified by a recent publication by Wang et al. (1989) using four structural techniques and involving six collaborators from four laboratories.

7. CONCLUSIONS

The basic structural unit within all carbon materials is the layer based on the graphitic lattice. However, the variations in the degree of perfection, the extent and the stacking arrangements of these layers results in a multitude of carbon forms.

A continuous range of structures is found, from non-graphitizable, isotropic carbons to graphitizable, anisotropic carbons. However, the majority of carbonised materials can be classified as being near to one or other of the extremes.

Non-graphitizable carbons are formed from parent materials which do not pass through a fluid phase.e.g. wood. The resultant structure retains much of the original arrangement and, therefore, has poor parallel stacking of any lamellar molecules together with a high degree of cross-linkage. These materials often exhibit microporosity and surface areas above 1,000 $m^2 g^{-1}$.

Graphitizable carbons have passed through a fluid phase during pyrolysis. The mesophase (discotic nematic liquid crystal) development during the fluid phase generates large extents of parallel stacked molecules. These are necessary for subsequent graphitization. The variation in cokes is due the variation in reactivity of the molecules of the parent materials which controls the sizes and shapes of the intermediate liquid crystals (mesophase). Surface areas are often much less than 10 $m^2 g^{-1}$.

7.1 The diversity of carbon

Carbon is a simple element but ....

One branch of <u>chemistry</u> is devoted to its compounds !

One branch of **science** is devoted to the many forms of the element as a solid material.

The best of this is that although most carbon materials are grey or black to the naked eye and the uninitiated, a closer examination reveals the form, beauty and even colour of carbon science.

References

Amelinckx, S., Delavignette, P. and Heerschap, M., (1965). Dislocation and stacking faults in graphite. Chemistry and Physics of Carbon, Ed. P.L. Walker Jr, Marcel Dekker, New York, **1**, 1.

Auguie, D., Oberlin, M., Oberlin, A. and Hyvernat, P., (1980). Microtexture of mesophase studied by high resolution conventional transmission electron microscopy. <u>Carbon</u> **18**, 337.

Baker, R. T. K., (1982). The relationship between particle motion on a graphite surface and Tammann temperature. <u>J. Catalysis</u> **78**, 473.

Baker, R. T. K. and Harris, P. S., (1978). Formation of filamentous carbon. Chemistry and Physics of Carbon, Eds. P.L. Walker Jr. and P. A. Thrower, Marcel Dekker, New York, **14**, 83.

Bennett, S. C., Johnson, D. J. and Johnson, W., (1983). Strength-structure relationships in PAN-based carbon fibres. <u>J. Mat. Sci.</u> **18**, 3337.

Blayden, H. E., Gibson, J. and Riley, H. L., (1944). An X-ray study of the structure of coals, cokes and chars. <u>Proc. Conf. on Ultrafine Structure of Coals and Cokes</u>, BCURA, 1943, Leatherhead.

Boehm, H. P. and Coughlin, R. W., (1964). Enthalpy differences of hexagonal and rhombohedral graphite. <u>Carbon</u> **2**, 1.

Bourrat, X., Oberlin, A. and Escalier, J. C., (1986). Microtexture and structure in semi-cokes and cokes. Fuel **65**, 1490.

Bradley,J. R.,Chen, Y.-L. and Sterner, H. W., (1985). The struvture of carbon filaments and associated catalytic particles formed during pyrolysis of natural gas in steel tubes. Carbon **23**, 715.

Brooks, J. D. and Taylor, G. H., (1969). The formation of some graphitizing carbons. Chemistry and Physics of Carbon, Ed. P.L. Walker Jr., Marcel Dekker, New York, **4**, 243.

Brown, T. R., Kronberg, B. I. and Fyfe, W. S., (1981). Semi-quantitative ESCA examination of coal and coke Surfaces. Fuel **60**, 439.

Cameron, A. E. and Wichers, E., (1962). Report of the International Commission on Atomic Weights. J.A.C.S. **84**, 4175.

Clarke, L. J., (1985). Surface Crystallography, John Wiley and Sons, Chichester.

Cottinet, D., Couderc, P., Saint-Romain, J. L. and Dhamelincourt, P., (1988). Raman microprobe study of head-treated pitches. Carbon **26**, 339.

CRC, (1985). Handbook of Chemistry and Physics, 66th Edition.

Donnet, J-B. and Bansel, R. P., (1984). Carbon Fibers, Marcel Dekker, New York, U.S.A.

Donnet, J-B. and Voet, A., (1976). Carbon Black, Marcel Dekker, New York.

Dubinin, M. M. and Plavnick, G. M., (1968). Microporous structure of carbonaceous adsorbents. Carbon **6**, 183.

Forrest, R. A. and Marsh, H., (1977). Reflectance interference colours in optical microscopy of carbon. Carbon **15**, 348.

Franklin, R. E., (1951a). Structure of graphitic carbons. Acta Cryst. **4**, 253.

Franklin, R. E., (1951b). Crystallite growth in graphitizing and non-graphitizing carbons. Proc. Roy. Soc. **A209**, 196.

Given, P. and Marzec, A., (1988). Protons of differing rotational mobility in coals. Fuel **67**, 242.

Grint, A., Swietlik, U. and Marsh, H., (1979). Carbonization and liquid crystal (mesophase) development. Fuel **58**, 642.

Gundermann, K-D., Humhe, K.,Emrich, E. and Rollwage, U., (1987). Extended models of coal structures. 1987 International Conference on Coal Science, Elsevier, Amsterdam, p 49.

Hirsch, P .B., (1954). X-ray scattering from coals. Proc. Roy. Soc. **A226**, 143.

Honda, H., (1988). Carbonaceous mesophase: history and prospects. Carbon **26**, 139.

Honda, H. and Yamada, Y., (1973). Meso-carbon microbeads. Sekiyu Gakkai Shi, (J. Japan. Petrol. Inst.) **16**, 392.

ICCTC, (1982). First publication of 30 tentative definitions. Carbon **20**, 445.

ICCTC, (1983). First publication of further 24 tentative definitions. Carbon **21**, 517.

ICCTC, (1985). First publication of 14 further tentative definitions. Carbon **23**, 601.

ICCTC, (1986). First publication of 9 further tentative definitions. Carbon **24**, 246.

ICCTC, (1987a). First publication of 8 further tentative definitions and second publication of tentativedefinition of term 29. Carbon **25**, 317.

ICCTC, (1987b). First publication of five further tentative definitions. Carbon **25**, 449.

Janosi, A. and Stoeckli, H. F., (1979). Comparative study of gas adsorption and small angle X-ray scattering by active carbons, in relation to heterogeniety. Carbon **17**, 465.

Jeffrey, J.W., (1971). Methods in X-ray Crystallography, Academic Press, London, p 83.

Johnson, D. T., (1987). Structural studies of PAN-based carbon fibres Chemistry and Physics of Carbon, Ed. P. A. Thrower, Marcel Dekker, New York, **20**, 1.

Kawamura, K. and Bragg, R. H., (1986). Graphitization of pitch coke: changes in mean interlayer spacing, strain and weight. Carbon **24**, 301

Kodama, M., Esumi, K, Meguro, K and Honda, H., (1988). Adsorption of human serum globulin in mesocarbon microbeads. Carbon **26**, 777

Komiyama, M., Tsunoda, T and Oyino, Y., (1985). A novel method of examining carbonaceous deposits on spent catalyst surfaces. Carbon **23**, 613.

Konnert, J. H. and D'Antonio, P., (1983). Diffraction evidence for distorted graphite-like robbons in an activated carbon of very large surface area. Carbon **21**, 193.

Kozlowski, C. and Sherwood, P. M. A., (1986). X-ray photoelectron spectroscopic studies of carbon fibre surfaces. Carbon **24**, 357.

Latham, C. S. and Marsh, H., (1986). Anodes for aluminium production. Ext. Abs. CARBON '86., 4th International Carbon Conference, Baden-Baden, Arbeitskreis Kohlenstoff, DKG, p 802.

Lespade, P., Marchand, A., Couzi, M. and Cruege, F., (1984). Caracterisation de materiaux carbones par microspecrtmetric Raman. Carbon **22**, 375

Lewis, I. C. and Lewis, R. T., (1988). "Carbonaceous mesophase: history and prospects" - A reply. Carbon **26**, 751.

Mantell, C. L., (1968). Carbon and Graphite Handbook, Interscience, New York.

Markovic, V. and Marsh, H., (1983). Microscopic techniques to examine structure in anisotropic cokes. J. of Microscopy **132**, 342.

Marsh, H. and Crawford, D., (1984). Structure in graphitizable carbon from coal-tar pitch HTT 750-1148 K studied using high resolution electron microscopy. Carbon **22**, 413.

Marsh, H. and Griffiths, J., (1982). A high resolution electron microscopy study of graphitization of graphitizable carbons. Extended Abstracts, International Symposium on Carbon, Toyohashi, Japan, p 81.

Marsh, H. and Walker, P.L. Jr., (1979). The Formation of graphitizable carbons via mesophase: chemical and kinetic considerations. Chemistry and Physics of Carbon, Ed. P. L. Walker Jr., Marcel Dekker, New York, **15**, 229.

Marsh, H. Forrest, M., and Pacheco, L. A., (1981). Structure in metallurgical cokes and carbons as studied by etching with atomic oxygen and chromic acid. Fuel **60**, 423.

Mernagh, T.P., Cooney, R.P. and Johnson, R.A., (1984). Raman spectra of graphon carbon black. Carbon **22**, 39.

Messier, R., Spear, K. E., Badzian, A. R. and Roy, R., (1987). The quest for diamond coatings. J. Materials **39**, 8.

Moreland, A., Patrick, J. W. and Walker, A., (1988). Optical anisotropy in cokes from high-rank coals. Fuel **67**, 730.

Moore, H., (1981). Highly ordered pyrolytic graphite and its intercalation compounds. Chemistry and Physics of Carbon, Ed. P. L. Walker Jr. and P. A. Thrower, Marcel Dekker, New York, **17**, 233.

Mochida, I., Korai, Y., Fujitsu, H. and Hatano, H., (1985). Synthesis and characterisation of mesophase spheres from hydrogenated pyrene through two-stage heat treatment under gas flow, HighTemperature - High Pressure **17**, 581.

Oberlin, A., (1984). Carbonization and graphitization. Carbon **22**, 521.

Patrick, J. W., Reynolds, M. J. and Shaw, F. S., (1973). Development of optical anisotropy in vitrains during carbonization. Fuel **52**, 198.

Pauling, L., (1960). The Nature of the Chemical Bond, Cornell University Press, Ithaca, N.Y., 3rd Edition.

Rand, B., (1985). Carbon fibres from mesophase pitch. "Strong Fibres", Ed W. Watt and B. V. Perov, Handbook of Composites, Vol 1, Ser. Eds. A. Kelly and Yu N. Rabotnov, North-Holland, Amsterdam, p. 495.

Riley, H. L., (1947). Amorphous carbon and graphite. Quart. Rev. Chem. Soc. **1**, 59.

Shimokawa, S., Yamada, E., Yokono, T., Yamada, J., Sanada, Y. and Inagaki, M., (1986). High-temperature high-pressure ^{1}H-nmr study on formation of carbon spherules. Carbon **24**, 771.

Shinn, J.H., (1984). From coal to single-stage and two-stage products: a reactivity model of coal structure. Fuel **63**, 1187.

Stach, (1982). Stach's Textbook of Coal Petrology, Gebruder Borntraeger, Berlin.

Wang, A., Dhamelincourt, P., Dubessy, J., Guerard, D., Landais, P. and Lelaurain, M. (1989). Characterization of graphite alteration in an uranium deposite by micro-Raman spectroscopy, X-ray diffraction, transmission electron microscopy and sacnning electron microscopy. Carbon **27**, 209.

Chapter 2

Mechanisms of Formation of Isotropic and Anisotropic Carbons

H. Marsh and R. Menendez*

Northern Carbon Research Laboratories, Dept. of Chemistry, University of Newcastle upon Tyne, Newcastle upon Tyne, NE1 7RU, U.K.

*Instituto Nacional del Carbon, Apartado 73, 33080 Oviedo, Spain.

Summary.

Structure in both isotropic (non-graphitizable) and anisotropic (graphitizable) carbons is based on the graphite molecule albeit in very defective forms. Isotropic carbons have lamellae arranged randomly with associations of small number to give an indication of short-range order. Anisotropic carbons have essentially approximately parallel arrangements of lamellae which subsequently assume increased order on heat treatment to give graphitic material. Isotropic carbons are generated from polymeric biomass materials, e.g. wood and synthetic resins which maintain their polymeric character on pyrolysis, without fusing, the carbon being a pseudomorph of the parent substance. Anisotropic carbons are generated from pitch and coals. These materials fuse on pyrolysis and it is from within this liquid phase that polynuclear, aromatic, discotic, nematic liquid crystals (mesophase) are generated by a process of homogeneous nucleation. Mesophase is the precursor to anisotropic structure in carbons/cokes. Pyrolysis chemistry is discussed indicating how optical texture of coal is a function of the reactivity of pyrolyzing systems. The importance of viscosity/temperature curves is described. Coal carbonization is more complex than pitch pyrolysis because of the need to depolymerise the original coal structure before liquid crystals can be generated from within the fluid coal.

MECHANISMS OF FORMATION OF ISOTROPIC AND ANISOTROPIC CARBONS

H. Marsh and R. Menendez

Northern Carbon Research Laboratories, Department of Chemistry, University of Newcastle upon Tyne, Newcastle upon Tyne, NE1 7RU, U.K.

1 INTRODUCTION

Carbon forms, except diamond, are based on the graphite lattice. Chapter 1, Figure 3, shows the structure of hexagonal graphite with layers or lamellae of carbon atoms arranged parallel to each other and lying above each other with an AB AB AB sequence. Carbon forms can be derived from such a model which is anisotropic, i.e. physical and chemical properties differ according to the axis selected: chemical bonding exists within the lamellae, but only van der Waals forces exist between the lamellae. Electrical conductivity is much greater parallel to the lamellae than across the lamellae.

Carbon forms are made up of such lamellae but with essential differences. First, the lamellae are much smaller in size than in single crystal graphite, in the range of about 5 to 500 nm. Second, the lamellae are not planar but can be bent and twisted. Third, the arrangement of carbon atoms within the lamellae is not perfect, holes are present which result in displacements from the true lattice configuration and there are also heteroatoms; hydrogen, oxygen, nitrogen, sulphur, bonded into the lamellae.

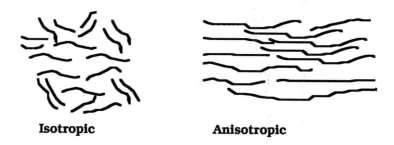

Figure 1. Two-dimensional drawings of carbon lamellae to illustrate structure in isotropic and anisotropic carbons.

The structure in many carbon forms gives the appearance of an assembly of small pieces of graphite lamellae, bent, twisted and imperfect such that the orientations of the lamellae are quite random but with, perhaps, some parallel stacking over short-range, Figure 1.

The bulk properties of these carbons, having an almost random arrangement of packed lamellae, are isotropic. The irregular stacking arrangements create space between the lamellae or groups of lamellae. This space constitutes microporosity. Differences in size and the defect nature of the lamellae create different packing sequences and, hence, differences in the properties of microporosity (Chapter 5). Surface areas are most frequently in the range 500 to 1,500 m^2g^{-1}. These carbons, which are isotropic, microporous and non-graphitizable, are formed from parent materials which do not fuse or melt during the pyrolysis/heat treatment/carbonization process. Such parent materials are often wood, non-fusing coals such as lignites and anthracites, nuts and nut shells and, in fact, most forms of biomass material.

If the lamellae are on average somewhat larger than previously described and less defective, with fewer heteroatoms, then the stacking arrangement is much more parallel. The degree of parallel stacking is such that the layers can be made significantly less defective when becoming graphitic on heat treatment to >2,500°C. Carbons prepared at, *e.g.* 1000°C would already be anisotropic and, hence, would be graphitizable. As the density of packing of larger lamellae is higher than for the smaller, more defective lamellae, then the inter-lamellar spacing is reduced to such an extent that the microporosity disappears. These anisotropic, non graphitic but graphitizable carbons are of low surface area, *e.g.* <10 m^2g^{-1}.

Parent materials for graphitizable carbons are dominantly petroleum pitch, coal-tar pitch, some coals, polynuclear aromatic systems and polyvinylchloride, all of which pass through a fluid phase during the pyrolysis/heat treatment /carbonization process. The detail of this process forms the basis of this Chapter.

2 ISOTROPIC CARBON

Isotropic, non-graphitizable carbons originate from materials which are already macromolecular in nature, *e.g.* the cellulose or lignin components of wood, nuts and nutshells, or the specific cross-linked structures (C-O-C bondings) of low rank coals such as peats, lignites, brown coals and non-caking bituminous coals. Synthetic resins of the chemical industry such as phenolic resin, polyfurfuryl alcohol and polyvinylidene chloride are of a similar nature.

To prepare an isotropic carbon, the precursor material must be polymeric, either being heavily cross-linked initially or developing cross-linkages in the early stages of carbonization. There can be fusion/melting of a macromolecular system (polymer) as in bituminous caking coals with subsequent development of macroporous chars. Otherwise, carbonization results in total volatilization as with linear polystyrene which does not give a carbonaceous residue.

The early stages of carbonization involve the cleavage of bonds within the macromolecular system to give free radicals. Electron spin resonance (ESR) studies (Mrozowski, 1982) show a rapid increase in free radical concentration above a heat treatment temperature of 500°C. Initially, as the carbon lattice develops, ESR studies indicate the presence of localized spin centres and conduction electrons. With higher heat treatment temperatures (HTT), the former are replaced by internal lattice spin centres associated with interstitial atoms, lattice vacancies and other imperfections. The localized spin centres, if accessible to air, can be eliminated by the paramagnetic oxygen molecules. Air sensitivity of signal is, thus, a monitor of microporosity which allows air to penetrate into the carbon.

Details of the collapse of macromolecular systems are not fully understood although the early stages of carbonization involve the elimination of small molecules as volatile material in the form of water, methanol, methane and carbon dioxide. Such eliminations create space or microporosity within the rigid macromolecular system and, at the same time, the radicals generated at surfaces either combine with each other or abstract hydrogen from the system.

The processes begin probably at temperatures as low as 350°C under nitrogen. Fourier Transform Infra-Red (FTIR) is probably the most suitable technique to study these systems and indicates that for a Phenadur resin, aromatic carbon of lamellar structure can be detected in materials of HTT as low as 400°C (Figure 2). Aliphatic side-chains are linked to the developing aromatic domains by mixed ketonic and ether linkages. The carbonization process is, thus, the simultaneous process of elimination of small molecules and the subsequent re-arrangements of carbon atoms (radicals) to form more stable six-membered rings of carbon lamellae, albeit imperfectly. At about 555°C, dependent on the system, all of the aliphatic carbon is converted to aromatic C-H. At about 730°C, any residual surface structures are no longer detectable by FTIR and at about 800°C, the specimen is no longer transparent to IR radiation and a carbon has been formed. The structure generated is isotropic as shown in Figure 1.

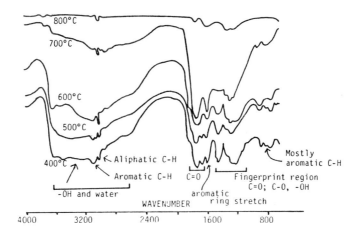

Figure 2. FTIR spectra generated from Phenadur resin carbonized to 800°C.

Carbons are obtained from materials such as cellulose and resins because a cross-linked polymer already exists in the parent material. A linear polymer, such as polystyrene (Neely 1981), produces no carbonaceous residue when pyrolysed, the process being one of depolymerization and chain scission. Cross-linked polystyrenes readily convert to carbons because of associated condensation reactions. Further, by introducing functionality following sulphonation, oxidation and chlorination, yields of carbons can be enhanced. The polystyrene system undergoes aromatization giving C/H atomic ratios corresponding to coronene with significant decreases in electrical conductivity (5 orders of magnitude, 600° to 800°C). This effect is very characteristic of the formation of carbon from an organic polymeric system. Much of the detail of the pyrolysis chemistry of various systems studied is reviewed by Fitzer et al. (1971).

A polymeric material that has attained significant application is polyacrylonitrile or PAN, used for the production of carbon-fibres. PAN fibres, to date, dominate the market in terms of tonnage applications. It is a matter of good fortune, in the comparison of textile fibres, that the pyrolysis of PAN proceeds via condensation reactions leading to a carbon rather than by depolymerization giving only liquid/gaseous products. PAN fibres are discussed further in Chapter 6.

In the thermal decomposition of PAN, nitrile groups (A) form tetrahydronaphthiridine rings (B) which aromatize to a more stable system (C), Figure 3.

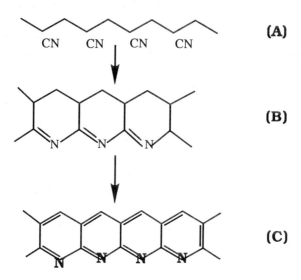

Figure 3. Cyclization of polyacrylonitrile.

It is generally considered that cyclization and dehydrogenation occur simultaneously. The cyclization of PAN is promoted by oxygen present in the gas phase during pyrolysis. At the back-bone of the heteroaromatic ladder polymers, oxygen uptake occurs e.g:-

The presence of oxygen affects aromatization processes in two ways. First, oxygen is an initiator for the formation of activated centres for cyclization and is responsible for their increased number. Secondly, the consecutive cyclization of these centres is inhibited by oxygen because of enhanced activation energy.

The stages of cyclization and aromatization, followed by growth of the aromatic clusters into the lamellae of carbon structure, lead to the three-dimensional structures so characteristic of carbon fibres as illustrated in Chapter 6.

The growth into the lamellae does not proceed uniformly and completely to give the graphitic structure, rather, defects within the lamellae are introduced and these persist into high temperature structures. The growth process varies from material to material and the pre-orientations within PAN fibre lead to carbon lamellae networks which optimise PAN in terms of mechanical properties (Chapter 6) and are responsible for its successful application.

Several techniques available for the analyses of structure within isotropic, non-graphitizable carbons characterize individual aspects but not the entirety of structure. Each parent material pyrolysed and the conditions of pyrolysis yield an individual carbon. The determination of adsorption isotherms (Chapter 5) provides semi-quantitative information about the sizes of spaces (porosity) between the lamellae of structure. X-ray and electron diffraction provide diffraction rings which need a model for interpretation yielding parameters for the elusive hypothetical crystallite (Chapter 1). It is quite easy to confuse size of lamellae with perfection of packing of lamellae, Figure 4.

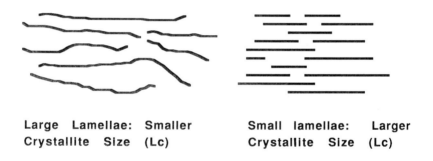

Large Lamellae: Smaller Crystallite Size (Lc)   **Small lamellae: Larger Crystallite Size (Lc)**

Figure 4. Two-dimensional drawings of carbon lamellae showing differences in relative alignments.

An extremely informative technique is phase contrast high resolution transmission microscopy (fringe imaging) which gives a two-dimensional aspect of structure in carbon. This, in association with light and dark field microscopy, provides adequate information to create an approximate model. However, this approach requires a well-equipped laboratory and experienced staff.

Oberlin (1984) presents a review of recent electron microscope observations relating to the carbonization and graphitization of many carbonaceous precursors. A very important conclusion emerges from this review. Of the many carbons studied, it is relevant to note that there is no break in the continuity of possible structures from isotropic carbons, of small lamellae size with short range order, to well-structured anisotropic graphitizable carbons. In terms of carbon usage, the extremes of the structure range have major industrial application, i.e.

isotropic active carbons used for purification and separation and highly graphitizable carbons such as regular- and needle-coke used in electrode manufacture.

Whereas the arrangement of lamellae (molecules) within isotropic carbons can be described by Figure 1, heat treatment to about 2800°C creates other structures which may possibly be dormant within the carbon. Figure 5 is a high resolution fringe-image electron transmission micrograph of structures found in carbon from polyvinylidene chloride ($(-CH_2-CCl_2)_n-$). In this figure, what appear to be ribbons of stacked lamellae are, in fact, the walls of cavities within the carbon, these cavities being about 5 to 10 nm (50 to 100 Å) in diameter. Similar structures are found in carbons from polyfurfuryl alcohol, other resins and carbonized saccharose which, although passing through a fluid phase (caramel), yield an isotropic, non-graphitizable carbon. Although earlier models based on these micrographs discuss structure in terms of three-dimensional entanglements of ribbons of structure, it is now considered that this model is incorrect and the cavity-wall model is favoured (Oberlin 1984). Figure 5 clearly shows the quite random orientation of the lamellae within the carbon over distances >20 nm. This randomness precludes any possibility of long-range graphitizability, although the crystallites of length (L_a dimension) ~10 nm and several lamellae thick (L_c dimension) may become graphitic. Thus, in discussing graphitizability within carbons, it is important to indicate the size of the graphitizable˙ structural units to prevent ambiguity and confusion.

The specificity of lamellar size, defect nature and stacking sequences critically controls the microporosity of the resultant carbon and this, in turn, controls the suitability for activation (enhanced porosity) of such carbons.

Another method for the formation of isotropic, non-graphitizable porous carbons is not regular carbonization but co-carbonization of coal/coke with potassium salts followed by leaching out of the potassium from the system (Marsh et al. 1982; Marsh et al. 1984). Although the mechanism is not clearly elucidated, it is most probable that potassium enters between the lamellae to form intercalation compounds. At higher temperatures within the process, the removal of potassium causes an exfoliation or separation of the lamellae. On subsequent removal of the potassium, these separated lamellae are grouped together again but with quite a different arrangement, more of a cage-like structure as indicated in Figure 6. The walls of the cage-like structures are made up of one or two lamellae only, unlike those of Figure 5.

Figure 5. Lattice fringe images of lamellae within an isotropic carbon, HTT 2700°C (2975 K) prepared from polyvinylidene chloride.

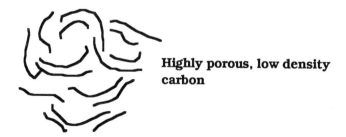

Figure 6. Diagram of lamellar arrangements in active carbons prepared from carbon and potassium carbonate (Marsh et al.1982).

The cage-like structures adsorb as if they were super-micropores and desorb without hysteresis. Effective surface areas can be as high as 6,000 m^2g^{-1} (Marsh et al. 1982). Adsorption within these cage-like structures and their possible expansion can account for high effective surface areas, far above the generally accepted levels of 1000-1500 m^2g^{-1}.

3   GRAPHITIZABLE CARBON - The problem

It is only within the last two decades that a realistic understanding of structure in non-graphitizable carbons has been developed. Similarly, despite the technological knowledge of synthesizing graphites via the Acheson process, it is again within the last two decades that the mechanisms of formation of anisotropic carbon have been elucidated.

Anisotropic, graphitizable carbons are produced from coal-tar pitch and petroleum pitch sources. These materials are complex mixtures of polynuclear hydrocarbon systems with functionalities and would appear to be ideal precursors for graphitizing carbons. Chapter 3 summarises the viscosity properties of these pitch materials. Whereas at temperatures below ~200°C the properties of the liquid pitch are decidedly non-Newtonian, at temperatures above 400°C they approach Newtonian behaviour. In such liquids, any association or 'ordering' of molecules is of very short-range order. In fact, at all levels of order, the bulk behaviour is that of an isotropic liquid, yet, over a very narrow temperature range, these pitches solidify to a "crystallite" coke which manifests various levels of long-range order, resulting in various levels of graphitizability.

Optical microscopy of polished surfaces of cokes (Chapter 1) indicates sizes of anisotropic areas (crystals almost) varying from <1.0 to about 250 µm. The larger the area, the higher is the degree of graphitizability. Not only do the pitch materials 'crystallize' into a solid coke but the cokes exhibit a wide range of crystallinity. The problem is the identification of mechanisms of formation of anisotropic carbons when taking into account the major differences of anisotropic area, mosaics to domains, in terms of pyrolysis chemistry.

3.1   Background

In 1944, Blayden et al. published a major study, using X-ray diffraction, of the carbonization of many precursors and established from line-broadening measurements that carbons from coals could be divided primarily into isotropic and anisotropic carbons. This was later taken up by Rosalind Franklin (1951) and further clarified. Blayden et al. (1944) wrote in 1944 that the *stacking of the lamellae in coal involves the movement of relatively large cumbersome units affected by the chemical and physical forces between the lamellae and their steric environment.* Later studies have confirmed this understanding in more precise detail. Franklin (1951), from X-ray studies, classified carbons (HTT >1000°C) into non-graphitizable and graphitizable, Figure 7. For carbons of HTT <1273 K (<1000°C), however, X-ray diffraction does not distinguish between carbons, phase contrast electron microscopy being a more suitable technique for structural analysis.

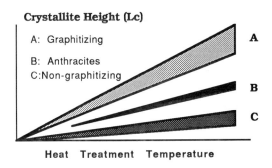

Figure 7. The general form of the crystallite height (L_C) as number of layers, for carbons of HTT 1273 K (Franklin 1951).

In 1964, in the University of Hull, Dr. J.J. Kipling carbonized polyvinylchloride ($-(CHCl)_n-$) which, unlike the similar polyvinylidene chloride ($-(CH_2-CCl_2)_n-$), gives an excellent graphitizable carbon. The obvious necessity of a fluid phase was recognised but not all fluid carbonizations gave anisotropic carbons. Additions of sulphur could destroy the anisotropy.

3.2 Mesophase - Early recognition

In Wongawilli in New South Wales, Australia, several coal seams have been metamorphosed via dyke and sill movement. Flows of magma either into the coal seam or parallel to it, temperatures of which are >1000°C, result in slow pyrolysis or carbonization of the coal over time-scales of years. In 1961, Taylor sampled and examined coals from metamorphosed seams. The effect over some hundreds of metres from the dyke was the establishment of a very gradual thermal gradient reaching towards the dyke to maximum temperatures which passed through and exceeded the temperatures of formation of anisotropic coke from coals. The slow rate of heating which had occurred and the very narrow temperature zone of the formation of anisotropic carbon provided the first clue as to the nature of the property of graphitizing ability. In previous laboratory experiments, because too rapid heating rates had been employed, the detail of the formation of anisotropic carbon had not been identified.

In coal samples approaching the dyke, Taylor (1961) noticed that the original bedding anisotropy of the coal (vitrinite) was lost and that there were small spheres of micron size of anisotropic material of higher reflectivity, when viewed in polished sections using optical microscopy with polarized light. Taylor recognised that the anisotropic spheres provided the clue to the development of graphitizable carbons. The molecular arrangements within these spheres, of equatorial parallel stacking rather than of circumferential stacking, Figure 8, closely resembled those of the nematic phase in substances which form LIQUID CRYSTALS. Thus, the mechanism of the formation of graphitizable carbons came to be closely associated with those of liquid crystals.

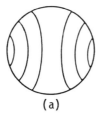 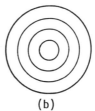

(a)   (b)

Figure 8.   Structure in mesophase: (a) the most common Brooks and Taylor structure of equatorial stacking of lamellae and (b) the less common onion structure of lamellae parallel to the external spherical surface.

3.3   Nematic liquid crystals

Liquid crystal systems have been recognized since 1888 although they were often just a curiosity in the laboratory. In 1922, it was suggested that the term "mesophase" (intermediate state) would obviate the apparent inconsistency of a 'crystalline liquid'. Although the term 'mesophase' is rather imprecise, it is currently in use and is discussed below.

3.4   Structure in liquid crystals

In 'normal' Newtonian liquids, only short-range order is to be found and that is at temperatures near to melting points, otherwise, the arrangement of molecules in space becomes more random. There is a range of organic molecules, rod-like in shape, with aromatic rings and dipole moments associated with functionality

(Figure 9). Bonding between such molecules, at temperatures near to melting points, is such that considerable ordering is maintained even in the fluid phase. The higher the 'cohesion energy' between the rod-like molecules, the more ordered is the liquid crystal phase. Of these molecular-types, some pass from the solid phase, on heating, to the smectic phase and then to the nematic phase (of less order), Figure 9. Others pass directly from the solid phase to the nematic phase, both ultimately becoming isotropic with further increase in temperature, i.e. they exhibit thermotropic properties.

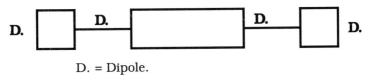

D. = Dipole.

(a) Rod-like molecules which form liquid crystals

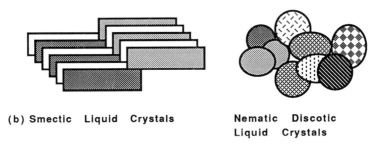

(b) Smectic Liquid Crystals   Nematic Discotic Liquid Crystals

Figure 9 (a) Diagram of structure in rod-like molecules which form liquid crystals. (b) Structure in smectic and nematic liquid crystals.

The transition temperatures are sharp and reproducible. The stability of the phases is totally dependent upon a balance between the cohesion forces between the molecules and the translational energy available to them at a given temperature. For Newtonian liquids, the translational energy exceeds the cohesion energy and association does not occur (although water may not fit this definition exactly). Another shape of molecule will also form liquid crystals, i.e. the disc-like or discotic molecule rather than the rod-like molecule. Truxene is a molecule which forms nematic liquid crystals and the discotic, aromatic condensed systems of products of pitch carbonization can form nematic liquid crystals, Figure 10 (a) (b).

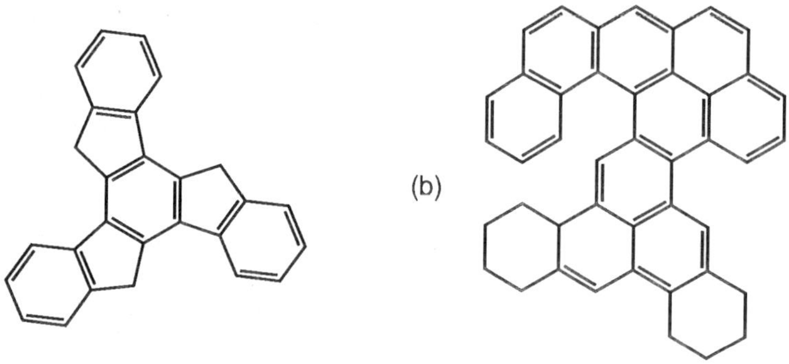

(a) α – Truxene

(b) Lamellar molecule of pitch pyrolysis

Figure 10 (a) Molecular formula of "α-Truxene" and (b) a possible structure in a lamellar molecule from pitch pyrolysis, both being discotic mesogens.

3.5 Nucleation of mesophase

Mesophase is not 'precipitated' from pitch solution, rather it is the growth of a new phase (as in the Wongawilli coal seam) by homogeneous nucleation. It does not occur because of insolubility of larger molecules within the pitch but because there is a preference for the formation of the liquid crystal phase with its higher stability. Mesophase growth cannot be 'seeded' as can solid crystals from supersaturated solutions. Attempts have been made to 'seed' mesophase growth by carbon blacks but without success.

Unlike the 'classical' liquid crystals of rod-like shape which are now used extensively in display systems (i.e. in chemically stable systems), the liquid crystals of pitch and coal carbonization are generated within the pyrolysis process itself and never reach an equilibrium position. The system is dynamically active and, it must be stressed, it is the variation in pitch pyrolysis chemistry which dictates properties of final cokes/carbons. Initially, the average molecular weight of pitch at 200°C is about 200 amu. Here, translational energies exceed cohesion energies and the fluid is approaching Newtonian and is isotropic. Pitch pyrolysis chemistry is, however, dominantly dehydrogenative polymerization and, at temperatures >400°C, molecular weights reach 600-900 amu when cohesive energy exceeds translational energy (Marsh and Menendez 1988). At this stage, the molecules remain attached to each other (initially by dispersion forces) following a collision. They collect further mesogen molecules of similar size and shape until, ultimately, clusters can be observed by the optical microscope of anisotropic reflecting spheres of 1 μm diameter. Pitch pyrolysis chemistry continues to play a part because the thermotropic properties of 'classical' liquid crystals are not usually observed. This is because some degree of polymerization continues within the mesophase such that cohesion energies are always ahead of translational energies. Of course, chemical cross-linkage between constituent mesogens of the clusters also thermally stabilizes the clusters against dissociation. By careful control of a temperature jump type of experiment (i.e. +5°C), the thermotropic properties of liquid crystals from pitch can be demonstrated, i.e. the system becomes isotropic but, on cooling, returns to being anisotropic. Greinke and Singer (1988) report molecular weight changes during stages of pitch pyrolysis and mesophase growth.

3.6 Structure within mesophase

During the initial growth stages of mesophase, the discotic mesogen molecules are stacked parallel and vertical to each other (Oberlin 1984). However, because of the dictates of minimum surface energy, the growing mesophase adopts a spherical shape. In some systems, such as coal or very reactive pitches, the growth units may never be of sufficiently low viscosity (high fluidity) to assume a spherical shape but remain irregular. Structure within the growth sphere is as described in Figures 8 and 11. The constituent lamellar molecules of the sphere are parallel to an equatorial plane as indicated by electron diffraction, (Brooks and Taylor 1968). Optical microscopy indicates that the layers become orientated towards the poles of the spheres and that they approach the surface of the sphere at right angles, as indicated in Figure 8(a). An alternative structure, the onion-shaped structure, Figure 8(b), has been reported and is observed when a sphere of mesophase is trapped within the porosity of a coke or in a second immiscible fluid.

The nematic liquid crystal structure depicted in Figure 8(a) is clearly confirmed by phase contrast high resolution electron microscopy (HREM) studies of mesophase and graphitizable carbon (Oberlin 1984). Marsh and Crawford (1984) studied changes in structure induced by heat treatment, 477° to 875°C. The process of graphitization, from the nematic stacking of mesophase to structures of graphitic carbon, is modelled in Figure 12. The HREM studies, by fringe imaging of the constituent lamellae, confirm the nematic stacking sequence of molecules in the mesophase as in Figure 13. A diagram modelling mesogen molecules and structure in mesophase spheres is in Figure 14 with Figure 15 being a SEM micrograph of clustered mesophase spheres from acenaphthylene (Marsh et al. 1973).

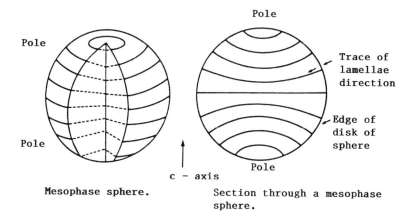

Figure 11. Structure of anisotropic nematic liquid crystals as spheres of mesophase within pyrolyzing isotropic pitch.

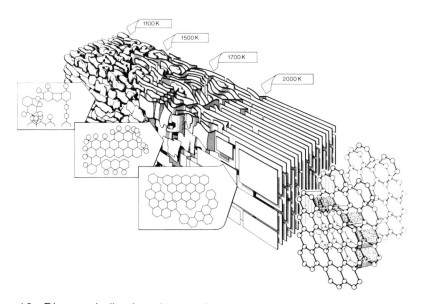

Figure 12 Diagram indicating changes in structure as non-graphitic mesophase is converted on heat treatment to graphitizable carbon. (Marsh and Griffiths, 1982).

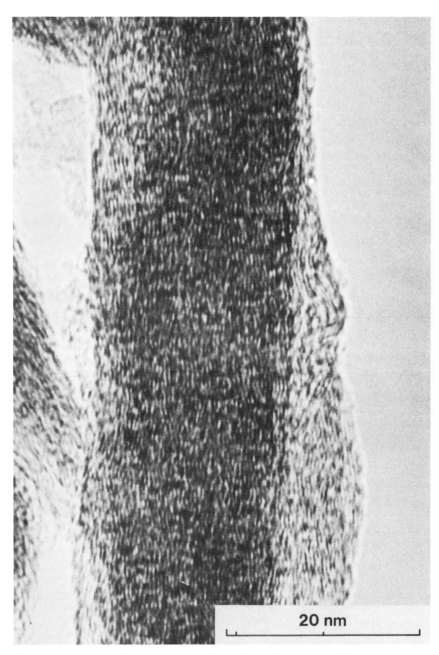

Figure 13. Lattice fringe images of lamellae within a graphitizing carbon, HTT 900°C, 1173 K, prepared from polyvinyl chloride.

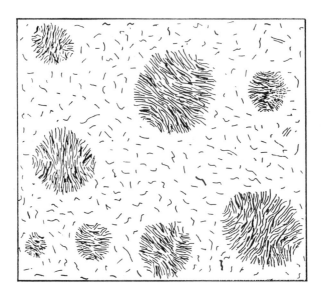

Figure 14. Diagram of mesogens within the matrix and mesophase spheres during pitch carbonization.

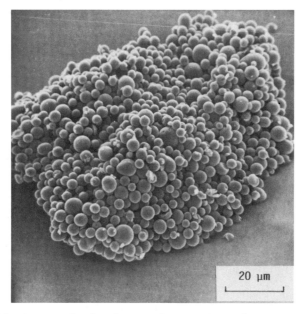

Figure 15. SEM micrograph of spheres of mesophase from acenaphthylene, HTT 550°C (Marsh et al. 1973).

4 CHEMISTRY AND VISCOSITY OF PYROLYSIS SYSTEMS

It must be stressed, initially, that the size of liquid crystals, i.e. as structure or optical texture observed by optical microscopy of cokes, relates immediately to the properties of mesophase at the time of solidification and these, in turn, are dependent upon the chemical properties of the parent pitch or coal. For example, it is not possible to increase the size of an optical texture (except within narrow limits) by physical manipulation of mesophase. This is not a means of coke quality improvement. Rather, improvements are brought about by 'improvements' to the quality of the parent feedstock. Once a feedstock is selected, then the optical texture of the resultant coke is more or less predetermined.

The generation of mesophase within a pyrolysing system only occurs if several constraints are operating, e.g:-

1. The intermolecular reactivity of constituent pyrolysis molecules is constrained to limited molecular growth of polynuclear aromatic (PNA) molecules, not exceeding ~900 amu on average (Greinke and Singer 1988).
2. The system must remain fluid to temperatures of 400° to 450°C such that the PNA molecules have sufficient mobility within the pyrolysing pitch to establish the liquid crystal system.
3. The intermolecular reactivity of constituent molecules of mesophase is also constrained to facilitate growth, coalescence and movement within the fluid liquid crystal system prior to solidification.

If the intermolecular reactivity of constituent pyrolysis molecules is too high, then the system will cross-link at temperatures below those at which mesophase is generated and the system solidifies (prematurely) to form an isotropic carbon (Marsh and Latham 1986, Mochida and Korai 1986).

Although the pyrolysis chemistry of pitch and coal is undoubtedly complicated, broad principles leading to control of mesophase are established. These can be summarized in discussion of Figures 16 and 17 which relate viscosity of pitches to heat treatment temperature and size of optical texture in resultant cokes to the temperature of minimum viscosity of the pitch system. The size and shape of optical texture as observed in a wide range of coke types varies from <1 µm to >200 µm. Metallurgical cokes and shot cokes have the smallest sizes and needle cokes have the largest sizes. A nomenclature to describe these textures, (Marsh 1982, Moreland et al. 1987 and Coin 1987), is in Table 1.

Differences between optical textures of cokes relate directly to differences in pyrolysis chemistry of the parent pitches. The broad chemical principles are as follows:-

(a) The largest sizes of optical texture (~200 μm) of coke result from highly aromatic parent materials, e.g. model compounds such as anthracene and acenaphthylene. Referring to Figure 16, the viscosity/HTT would include a minimum in viscosity of all pyrolysing systems, AB, followed by a necessary period of almost constant viscosity, BC, rise in viscosity as mesophase is formed, CD, eventually leading to solid green coke, DE.

(b) For pitch materials, other carbonization systems will produce optical textures of smaller size associated with higher (minimum) viscosities, PQ, with a shorter range of minimum viscosity and lower temperature of minimum viscosity, QR, and a rapid rise to solid coke, RS, Figure 16.

(c) The transition from largest textures (domains) to smallest textures (mosaics) is associated with enhanced intermolecular reactivity resulting from:
    (i)   less aromatic feedstocks
    (ii)   the presence of alkyl side-chains
    (iii) the presence of reactive functional groups, e.g. hydroxyl, carboxyl, etc.
    (iv) the presence of heteroatoms within the molecules.

(d) As intermolecular reactivity is increased, so the transition towards isotropic carbon proceeds continuously with decreasing size of optical texture until it cannot be detected by optical microscopy.

(e) The transition to decreasing size of optical texture is associated with increasing activity of transient free radicals (Yokono et al. 1966). If the transient free radicals can be stabilized to promote intermolecular stability (to (a) as above) then sizes of resultant optical textures can be enhanced. The concept of transferable hydrogen applied to mesophase growth (Marsh and Neavel 1980) has proved successful and has afforded a means of improving feedstocks for coking. In fact, the role of hydrogen transfer reactions and the availability of transferable hydrogen dominates pyrolysis chemistry leading to mesophase, as summarized by Yokono et al. (1966), see Figures 18, 19.

Table 1. Nomenclature to describe size and shape of optical texture in anisotropic cokes, using polished surfaces.

Isotropic	(I)	No optical activity
Very fine-grained mosaics	(VMF)	<0.5 µm in diameter
Fine-grained mosaics	(Mf)	<1.5 > 0.5 µm in diameter
Medium-grained mosaics	(Mc)	<10.0 > 5.0 µm in diameter
Supra mosaics	(SM)	Mosaics of anisotropic carbon orientated in the same direction to give a mosaic area of isochromatic colour.
Medium-flow anisotropy elongated	(MFA)	<30 µm in length; <5 µm in width
Coarse-flow anisotropy elongated	(CF)	<60 > 30 µm in length; <10 > 5 µm in width
Acicular flow domain anisotropy	(AFD)	>60 µm in length; <50 µm in width
Flow domain anisotropy elongated	(FD)	
Small domains, ~ isochromatic	(SD)	
Domains, ~ isochromatic	(D)	<60 µm in diameter

D_b is from basic anisotropy of low-volatile coking vitrains and anthracite.

D_m is by growth of mesophase from fluid phase.

Ribbons (R) strands of mosaics inserted into an isotropic texture.

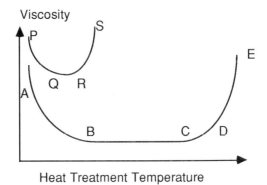

Figure 16. Pyrolysis of pitch: Variation of viscosity with heat treatment temperature.

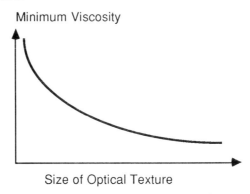

Figure 17. Pyrolysis of pitch: Variation of size of optical texture of coke with minimum viscosity.

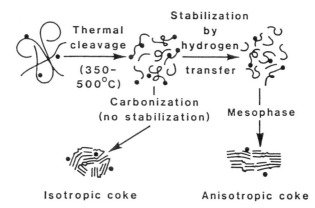

Figure 18. The influence of hydrogen transfer reactions during pyrolysis of petroleum residues on carbon formation (Yokono et al. 1986).

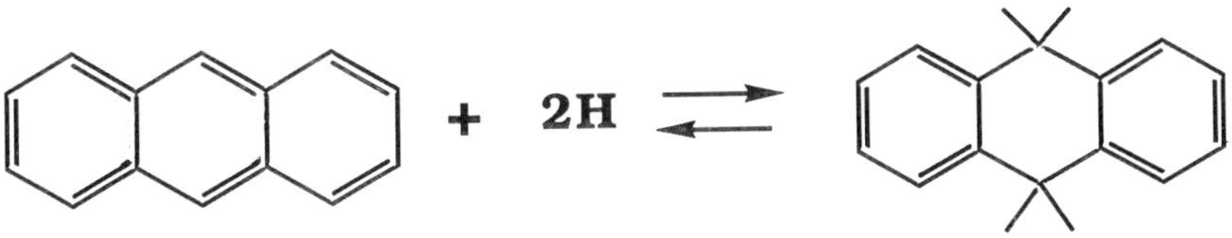

Anthracene + 2H ⇌ 9,10 Dihydroanthracene

Figure 19. Hydrogen transfer between anthracene and 9,10 -dihydroanthracene.

4.1 Growth and properties of mesophase: Summary

1. The growth of mesophase is essentially a delayed process.
2. Earlier condensations/polymerizations associated with functionality and radicals must be constrained as they lead to isotropic carbon.
3. The importance of HYDROGEN TRANSFER availability, to delay molecular growth, is to be noted.
4. It is necessary to form stable polynuclear aromatics ~900 amu in a low viscosity medium (HTT >400°C) to create mesophase, by HOMOGENEOUS NUCLEATION, i.e. a physical process.
5. Initially, the mesophase is thermotropic.
6. Although chemically stable (relatively) at the temperature of formation, increasing HTT promotes cross-linking into a macromolecular structure.
7. The mesophase has flow properties facilitating the introduction and annihilation of disclinations (see below).
8. The resultant size of optical texture in cokes controls such properties as (a) STRENGTH (b) THERMAL AND ELECTRICAL RESISTANCE (c) OXIDATION RESISTANCE and (d) THERMAL SHOCK RESISTANCE.

4.2 Aspects of mesophase chemistry

The physical properties of mesophase can be changed by chemical treatments, an approach very necessary to control mesophase viscosity as in the spinning of fibres. Some of these aspects are described by Mochida and Korai (1986). Hydrogenation enhances stability and, accordingly, lowers viscosity. Generally, the higher the heat treatment (HTT) of the mesophase, the more difficult it is to hydrogenate. Extraction (solubilization) of mesophase is of commercial interest. Alkylation enhances the solubility of mesophase but alkyl groups tend to be thermally unstable and contribute little to fusibility. Alkylation prior to hydrogenation, however, is effective in increasing the yield of mesophase.

Oxidative treatments remove hydrogen and introduce oxygen into the mesophase (Mochida et al. 1983) which promotes condensation and thermosetting. Naphthenic hydrogen is more reactive than aromatic hydrogen, (Mochida et al. 1982). Promotion of reactive grouping facilitates the thermosetting process of mesophase in fibres (see Chapter 6).

4.3 Mesophase growth and coalescence

So far, aspects of the development of mesophase have been considered. However, the coalescence of the growth spheres of mesophase into a single mesophase material, with the associated introduction and annihilation of defects called disclinations, is a major study area of the subject. The residual disclinations and eventual control of these disclinations are factors in the control of the mechanical properties of carbons derived from mesophase. The studies of White (1976) are significant in this area.

Figure 20 is an optical micrograph of the optical texture of a polished surface of mesophase derived from a petroleum pitch at 750 K. It shows a sphere of residual pitch within the fluid mesophase, indicative of the high fluidity of mesophase at 475°C. Within this pitch sphere are observed small spheres of mesophase being generated at the later stages of pyrolysis. The optical texture is indicative of flow domains (Table 1) of a highly fluid mesophase. The optical texture is not rigid but changes as the liquid crystal flows within the pyrolyzing system. From the extinction characteristics of the surface, using polarized light optical microscopy, it is possible to map out the emergence orientations at the surface of the lamellar molecules of the mesophase. White (1976) has made an extensive study of this approach to structural analysis and an example of the maps so generated is in Figure 21. The mesophase, as it flows, establishes and anniliates the 'disclinations' or gross defects within the liquid crystal system. A nomenclature of nodes and crosses to describe these disclinations is in Figure 21. In the manufacture of needle-cokes, the physical conditions in the delayed coker create needle-like structural units within the flowing mesophase.

However, not all mesophases have the same viscosities (Figure 16). For pyrolysis systems which produce high viscosity mesophase, the growth units of mesophase of small sizes of ~1-100 μm are so viscous that coalescence is not possible. They tend to adhere together to form mosaics, e.g. metallurgical cokes and shot cokes.

The coalescence of mesophase is very sensitive to the presence of particulate matter within the pyrolyzing system (Kuo et al. 1987). During the growth of mesophase from the pitch system, the particulate impurities (usually <1 μm dia.) adhere to the surface of the sphere. This prevents coalescence on touching (usually a very rapid process). Consequently, the viscosity of the mesophase increases with increasing time/temperature, this can never coalesce and smaller optical textures are generated. Coal-tar pitches can contain up to ~12 wt% of particulates (QI = quinoline insolubles). Petroleum derived pitches do not contain QI.

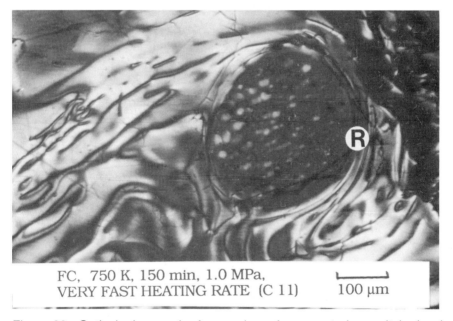

FC, 750 K, 150 min, 1.0 MPa, VERY FAST HEATING RATE (C 11)   100 μm

Figure 20. Optical micrograph of mesophase from a petroleum pitch showing a sphere of pitch (R) within the fluid mesophase (Romero 1989).

4.4 Carbons/cokes from mesophase from pitch

Cokes from petroleum and coal-tar pitch and from coals are of industrial importance. Cokes from pitch are most commonly prepared in a delayed coker and are produced in two commercial grades, regular coke for use in anode manufacture (aluminium pot cells) and premium needle-coke for electrode production (steel making). The delayed coker also produces shot coke (Marsh et al. 1985) which is of no commercial value.

Disclinations in Mesophase.

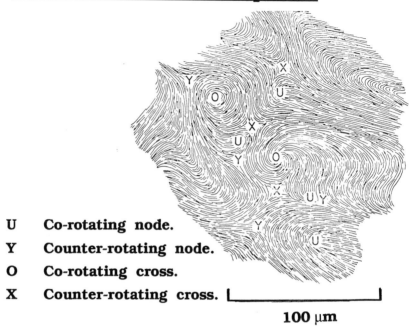

U   Co-rotating node.
Y   Counter-rotating node.
O   Co-rotating cross.
X   Counter-rotating cross.

100 μm

Figure 21. The lamelliform structure of coalesced mesophase derived from the optical texture.

The original design of the delayed coker converted the waste heavy liquid residues of petroleum refining into a solid 'coke' which was a more suitable material for disposal, either in land-filling or as a fuel. A later appreciation of the commercial value of suitable coke by-products stimulated the up-grading of pitch feedstocks being fed to the delayed coker. However, as the up-grading of low quality feedstocks proved to be difficult, considerable development of the process was undertaken in order to produce high quality delayed cokes from petroleum pitch feedstocks. The chemical composition of the parent pitch determines the quality of the final coke. The sequence of events can be summarised as follows:-

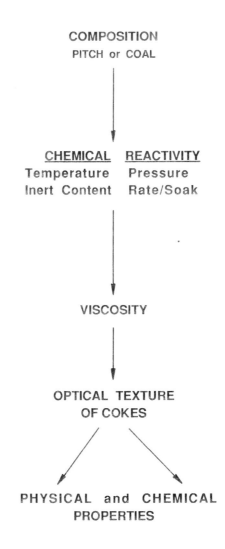

Of major importance is the 'reactivity' of the pitch during pyrolysis. Large size of optical texture requires chemical stability of molecular species to <400°C to maintain a liquid of relatively low viscosity so enabling flow and coalescence of mesophase to occur. If the reactivity is relatively higher, then extensive intermolecular growth occurs within the mesophase, viscosity rises and growth and coalescence is inhibited. Ultimately, reactive systems produce only isotropic carbon.

Improvements can be made to some feedstocks by blending (a method well established in coal carbonizations). The blending additive is usually a hydrogen donating agent which stabilizes the radicals of pyrolysis, as depicted in Figures 18, 19.

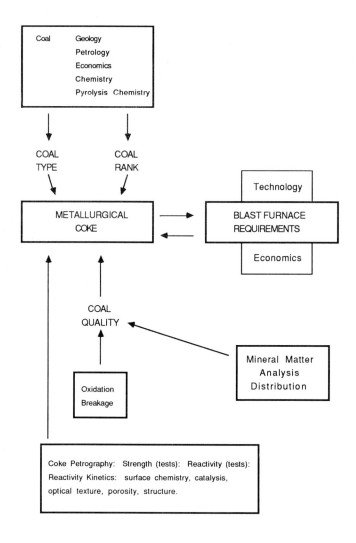

Figure 22. Summary of science and technology related to manufacture and use of metallurgical coke.

5. MESOPHASE FROM COAL
5.1 Metallurgical coke

The background to coke manufacture for coal is discussed in Chapters 8 and 9. Figure 22 summarises the coal science and technology relevant to the production of metallurgical coke. Coals purchased for use in carbonization blends must be characterized in terms of their RANK (defined as % oil reflectance of the vitrinite), their TYPE (defined in terms of maceral content) and QUALITY (mineral matter, oxidation).

5.2 Coal chemistry

To gain an understanding of the subject of coal chemistry and of coal carbonization, it should be noted that, unlike the pitch materials discussed above in terms of mesophase formation which are essentially liquids at temperatures >100°C, coal is a SOLID macromolecular, cross-linked material that has to undergo severe thermolysis or molecular degradation before a fluid phase is created. Not all coals produce a fluid phase. The fluidity of coals is not measured in units of viscosity (as with pitch) but mostly in units of dial divisions per minute (ddpm) of the Gieseler plastometer. The range of variation of maximum (ddpm) with rank is indicated in Figure 23, using volatile matter as the rank indicator.

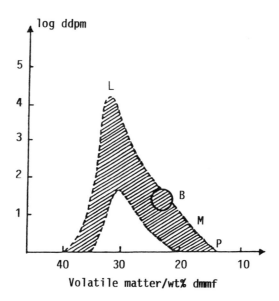

Figure 23. Diagram to indicate variations of maximum coal fluidity (ddpm:Gieseler) with rank (volatile matter).

As in the case of anthracites, the lowest rank bituminous (high volatile) brown coals and lignites do not fuse. This is due to extensive covalent linkages within the parent coal which prevent fusing (as in wood). With increasing rank, the oxygen-dominated cross-linkages diminish as oxygen is lost from the coal to be replaced by hydrogen-bonding arrangements within the macromolecular networks. Hydrogen-bonding is weaker than covalent bonding and the coals exhibit fluidity/plasticity on heating. With further increase in rank, this type of bonding is lost with the development of aromaticity and carbon-carbon cross-linkages producing anthracites which also have non-fusible structures. The wide variations in fluidities of coals (Figure 23) observed in the range 30-35 wt% volatile matter (dmmf), apart from maceral differences, are accounted for in terms of extents and strengths of cross-linkage by hydrogen bonding in the reactive components of the coal. The regular testing methods of coals do not indicate cross-link densities. From the viewpoint of carbonization of coking coal blends, this aspect of the macromolecular structure of coal has, to date, received very little attention. The world-wide trading of Southern and Northern Hemisphere coals now makes such studies of immediate relevance.

5.3 Mesophase formation during coal pyrolysis

Because coals are solid materials prior to carbonization, and a fluid system has to be created by prior thermolysis for mesophase formation, the type of diagram relating viscosity to HTT as for pitch (Figure 16) does NOT apply at all to coals. There are two factors which influence mechanisms of mesophase growth within fluid coals. Firstly, maximum coal fluidity developed during pyrolysis processes is rank dependent indicating wide differences in the average molecular weights of material in the fluid phase. Higher molecular weights prevent adequate movement within the fluid phase and, in addition, steric considerations are not conducive to liquid crystal growth. Secondly, coal rank dependency introduces wide differences in the chemical molecular composition of material in the fluid phase.

Coals of lowest rank which soften do not form anisotropic carbon because the viscosities of the system are not low enough to facilitate adequate molecular movement. Even with the least viscous (most fluid) of the coals (high volatile bituminous), oxygen functionality maintains a level of reactivity sufficiently high to prevent development of appropriate mesogens and isotropic carbon is created, despite the fluidity.

With increasing coal rank, i.e. decreasing volatile matter content (L to M, Figure 23) three changes occur relevant to mesophase development and growth: (a) the fluidity decreases (viscosity increases), this being adverse to mesophase generation: (b) the aromaticity increases and the molecular shape may become more planar, these being beneficial: (c) reactive functionality decreases, this being beneficial to mesophase generation.

As indicated above, discussions of the viscosity of pitch (and its rheological properties) are based on viscometer measurements, whilst the fluidity/plasticity of coal is measured with a Gieseler plastometer, units being the ddpm. For a mixture of mesophase and Ashland A240 pitch, viscosities are of the order of 10 Pa s. Balduhn and Fitzer (1980) published viscosity data for heat treated pitches, up to 500°C, viscosities rising from values of a few mPa s (350°C) to about 10 Pa s at 420°C. The rheology of the pitch systems appears to be of Newtonian behaviour, <300°C, and non-Newtonian, >300°C. Read et al. (1985) studied the rheological properties of three bituminous coals reporting minimum apparent viscosities between 7×10^3 to 10^5 Pa s (11,400 to 440 ddpm) at 420°-440°C. These viscosities are considerably higher than those of pitch materials. Other coals are reported to have minimum viscosities in the range 1.3×10^2 (mv B) to 4.7×10^7 Pa s (1v B). The rheological properties of coals are extremely complex and arise because coal melts consist of liquid, gaseous and solid phases. Certainly, in the temperature range 390°-450°C, coal melt is non-Newtonian. The degree of non-Newtonian behaviour is both time and temperature dependent such that it changes for a given coal (and from coal to coal) under isothermal conditions or under increasing temperature.

Thus, in the carbonization of coal blends which, industrially, form acceptable metallurgical cokes, a balance of properties results permitting the development of some mesophase, of small size and high viscosity, to create the mosaic structures of metallurgical coke. This balance of properties is a successful compromise and occurs at the same time as the volatile matter content is of the 'correct' magnitude to create the densities needed in cokes by expansion of a plastic layer which itself is of optimum properties. Other sizes of optical texture within a coke would be a disadvantage in terms of reactivity and strength.

Referring to Figure 23, increasing size of optical texture results in cokes from coals of increasing maximum viscosity (decreasing fluidity), Positions L to M, which is opposite to the conclusion from the study of pitches, Figures 16. With the highest ranks of coal, meta-anthracites and anthracites, decreasing fluidity is associated with covalent cross-link development and the facility for mesophase development is reduced, Figure 23, M to P. The anisotropy observed in carbons from anthracites originates as the basic anisotropy of the bedding properties of anthracites and is in no way related to mesophase growth.

Coal blends which provide cokes of adequate quality have fluidities in the region of about 150-200 ddpm as indicated by Position B of Figure 23. This is a compromise in properties: (a) the molecular thermolysis products in fluid coal (from the macromolecular systems) must have the required thermal stability, aromaticity and planarity to form liquid crystals, just as in pitch pyrolysis, (b) the fluidity (viscosity) is the minimum needed in the fluid coal for molecular movement to associate the molecules into liquid crystal clusters, (c) the viscosity of the mesophase is the minimum needed for surface adhesion of growth units but does not permit coalescence to the extents observed by pitch materials, (d) the temperature range of maximum fluidity is the widest generally observed in carbonizations of coals of different rank, (e) the maximum fluidity of the pyrolyzing coal is such that it permits porosity to develop from volatile release which is a working compromise between density and strength in the blast furnace, (f) the reactivities of coal molecules in this region are low enough to facilitate growth of liquid crystals but high enough to permit only small sized optical texture to develop, usually of mosaics, <10 μm diameter (Table 1).

Within recent years, attention has been given to the use of petroleum and coal-based pitch-like products to replace prime coking coals in blend formulations. A systematic study has been made of the science relevant to the use of pitch-like materials in blend formulations. Initially, using model compounds, Mochida and Marsh (1979) studied the co-carbonization of a range of coals with acenaphthylene and decacyclene, the latter compound successfully modifying the carbonization of coals of most ranks. Grint et al. (1979) extended the study by using Ashland A200 petroleum pitch as an additive and noted the ability of this pitch to interact with most coals. Similar effects, although not so extensive, were observed when solvent refined coals and coal-extracts were used as additives. Hydrogenated derivatives of petroleum pitches and coal extracts are superior modifiers to non-hydrogenated materials.

In an examination of the co-carbonization of one coal with several industrially supplied petroleum pitches, Grint and Marsh (1981) noted that the pitches differed considerably in their ability to modify the carbonization process of the coal. Some pitches (passive) had little effect on the coal, the product being a composite of clearly identifiable coal-coke and pitch-coke marked by sharp boundaries between the coke types. With active pitches, the latter interacted with the coal to produce a coke of different optical texture, no pitch-coke being detectable in the resultant composites and with diffuse boundaries of optical texture between the coke types. Grint and Marsh (1981) then demonstrated, with laboratory prepared cokes and with cokes prepared in a 7 kg oven, that a significant increase in coke strength was associated with the use of pitch additives in coal blends of marginal acceptance.

It is well established that drying of coals prior to charging to a coke oven and the technology of preheating of coal blends both result in improvements to coke quality. Carbonizing times are reduced: bulk density is increased to around 810 kg m^{-2} (Allen and Pettifor, 1984): a stronger coke results: coke oven operation is improved: the range of coals for coke making is widened to include poorer coking coals.

Over the years, it has proved difficult to identify the changes in coke structure brought about by preheating which result in improved coke strength. Microtextural analysis (of optical texture) could not be adequately quantified to distinguish cokes from wet and dry charging.

Figure 24 is a diagram of the cross-section (vertical) between walls of a coke oven charged with preheated coal showing the uniform and symmetrical nature of the plastic zone CC^1. However, during the carbonization of <u>wet</u> coal, the moisture leaves the coal by passing through the plastic zone, distorts the nature of this zone introducing asymmetry and gaps into the movements of the plastic zones towards the centre of the oven. Experiments described by Beck (1987) in which argon is introduced into the green coal show that argon does <u>not</u> diffuse through the plastic zone with preheated coal but does diffuse through the plastic zone of non-preheated coal (high V.M.). Thus, with preheating, the plastic zone can be considered to be impermeable.

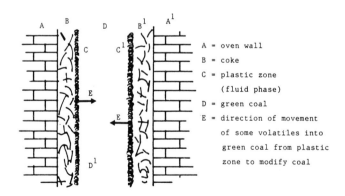

A = oven wall
B = coke
C = plastic zone (fluid phase)
D = green coal
E = direction of movement of some volatiles into green coal from plastic zone to modify coal

Figure 24. Diagram of cross-section (vertical) between walls of a coke oven charged with preheated coal showing uniform plastic zone, CC^1.

During the time that the plastic zone exists, coke structure is established. In reality, the carbonization of a single coal or of a coal blend can be considered as a co-carbonization of a coal with coal-tar (the relevance of pitch/coal co-carbonizations can be applied). Volatile matter release, particularly of aromatics, should not be considered only as a physical process. If the volatile matter can react chemically with the as yet uncarbonized coal, then this coal is modified prior to carbonization. The additional presence of hydrogen donating species from the volatile matter assists the breakage of the macromolecular structure of coal and promotes the stability of the molecular entities of the coal-melt. This process of auto-assistance within the coal system appears to be the process promoted by preheating. The release of moisture as water vapour through the plastic zone carries away volatile matter from the coal of the plastic zone to the outer gases. The relatively more quiescent conditions of preheating allow volatile matter to interact with green coal at the inner boundaries of the plastic zone so promoting the above effect. Thus, preheating is beneficial because (a) it improves the packing density of bulk coal in the oven (b) it permits an enhancement of 'Coking Capacity" within a given coal blend. No doubt, these effects alter the optical textures of resultant coke (wet v dry charging) so improving coke quality. Currently, there is a problem in obtaining statistically reliable quantitative data of optical texture sizes.

References

Allen, P.S. and Pettifor, S.I., (1984). Coal preheating - a practical application. Year Book, Coke Oven Managers' Association, Mexborough, U.K., pp. 136-160.

Beck, K.G., (1987). Blast Furnace Coke Production: The coking capacity determines the special process technology. Erdol und Kohle **40**, 129.

Blayden, H.E., Gibson, J. and Riley, H.L., (1944). An X-ray study of the structure of coals, cokes and chars. Proc. Conf. Ultrafine Structure of Coals and Cokes. , BCURA (London) p. 176.

Brooks, J.D. and Taylor, G.H., (1968). The formation of some graphitizing carbons. Chemistry and Physics of Carbon **4**, 243.

Balduhn, R. and Fitzer E., (1980). Rheological properties of pitches and bitumines up to temperatures of 500°C. Carbon **18**, 155.

Coin, C.A., (1987). Coke microtextural description: comparison of nomenclature, classification and methods. Fuel **66**, 702.

Fitzer, E., Mueller, K. and Schaefer, W., (1971). Conversion of organic compounds to carbon. Chemistry and Physics of Carbon **7**, 237.

Franklin, R., (1951). Crystallite growth in graphitizing and non-graphitizing carbons. Proc. Roy. Soc. A209, 196.

Greinke, R.A. and Singer, L.S., (1988), Constitution of coexisting phases in mesophase pitch during heat treatment: Mechanisms of mesophase formation. Carbon **26**, 665.

Grint, A. and Marsh, H., (1981). Co-carbonization of a high volatile caking coal with several petroleum pitches. Fuel **60**, 513.

Grint, A. and Marsh, H., (1981). The carbonization of coal-blends: Mesophase formation and coke properties. Fuel **60**, 1115.

Grint, A., Swietlik, U. and Marsh, H., (1979). The co-carbonization of vitrains with Ashland A240 petroleum pitch. Fuel **58**, 642.

Kipling, J.J., Sherwood, J.N., Shooter, P.V. and Thompson, N.R., (1964). Factors influencing the graphitization of polymer carbons. Carbon **1**, 315.

Kuo, K., Marsh, H., and Broughton, D.A., (1987). Influence of primary QI and particular matter on pitch carbonizations. Fuel **66**, 1544.

Marsh, H., (1982). Metallurgical coke, formation, structure and properties. Ironmaking Proceedings of the Iron and Steel Society of AIME, Warrendale, PA, USA, 41, p. 2.

Marsh, H. and Crawford, D., (1984). Structure in graphitizable carbons from coal-tar pitch, HTT 750-1148 K, studied using high resolution electron microscopy. Carbon **22**, 413.

Marsh, H., Crawford, D., O'Grady, T.M. and Wennerberg, A., (1982). Carbons of high surface area. A study by adsorption and high resolution electron microscopy. Carbon **20**, 419.

Marsh, H., Calvert, C. and Bacha, J., (1985). Structure and formation of shot coke - a microscopy study. J. Mat. Sci. **20**, 289.

Marsh, H., Dachille, F., Iley, M., Walker, P.L., Jr. and Whang, P.W. (1973). Carbonization and liquid-crystal (mesophase) development. Part 4, Fuel **52**, 253.

Marsh, H. and Griffiths, J., (1982). A high resolution electron-microscopy study of graphitization and graphitizable carbon. Int. Symp. on Carbon, Toyohashi, Japan, Kagaku Gijutsu-sha, Tokyo, 113.

Marsh, H. and Latham, C.S., (1986). The chemistry of mesophase formation. Petroleum Derived Carbons, ACS Symposium No. 303, Washington DC, U.S.A., p. 1.

Marsh, H. and Menendez, R., (1988). Carbons from pyrolysis of pitches, coals and their blends. Fuel Processing Technology 20, 269.

Marsh, H., and Neavel, R.C., (1980). A common stage in mechanisms of coal liquefaction and of carbonization of coal blends for coke making. Fuel 59, 511.

Marsh, H., Yan, D.S., O'Grady, T.M. and Wennerberg, A., (1984). Formation of active carbons from cokes using potassium hydroxide. Carbon 22, 603.

Mochida, I. and Korai, Y., (1986). Chemical characterization and preparation of the carbonaceous mesophase. Petroleum Derived Carbon, ACS Symposium No. 303, Washington D.C., U.S.A., p. 29.

Mochida, I. and Marsh, H., (1979). The co-carbonization of coals with acenaphthylene and decacyclene. Fuel 58, 633.

Mochida, I., Tamaru, K., Korai, Y., Fujitsu, H. and Takeshita, K., (1982). Carbonization properties of partially hydrogenated aromatic compounds-II. Carbon 20, 231.

Mochida, I., Tamaru, K., Korai, Y., Fujitsu, H. and Takeshita, K., (1983). Carbonization properties of hydrogenated aromatic hydrocarbons-III. Carbon 21, 535.

Moreland, A., Patrick, J.W., and Walker, A., (1987). Coke microtextural description. Fuel 66, 1310.

Mrozowski, S., (1982). ESR of carbons in the transition range. Carbon 20, 303.

Neely, J.W., (1981). Characterization of polymer carbons derived from porous sulphonated polystyrene. Carbon 19, 27.

Oberlin, A., (1984). Carbonization and graphitization. Carbon 22, 521.

Read, R.B., Ruecroft, P.J., Lloyd, W.G., (1985). Rheological properties of selected bituminous coals. Fuel **64**, 495.

Romero, E., (1989). Private communication.

Taylor, G.H., (1961). Development of optical properties of coke during carbonization. Fuel **40**, 465.

White, J.L., (1976). Mesophase mechanisms in the formation of the microstructure of petroleum coke. Petroleum Derived Carbons, ACS Symposium No. 21, Washington D.C., U.S.A. p. 282.

Yokono, T., Obara, T., Sanada, Y., Shimomura, S. and Imamura, T., (1986). Characterization of carbonization reaction of petroleum residues by means of high resolution ESR and transferable hydrogen. Carbon **24**, 29.

Chapter 3

Physical Properties of Pitch Relevant to the Fabrication of Carbon Materials

B. Rand, A.J. Hosty and S. West

Dept. of Ceramics, Glasses and Polymers, University of Sheffield, Elmfield, Northumberland Rd., Sheffield, S10 2TZ, U.K.

Summary.

The role of pitch as a binder in the fabrication of carbon materials is discussed.

Pitch is a mixture of many different molecules, some of which are mesogenic in character. The chemical composition, which is determined by the distillation and heat treatment conditions used in its production, controls the physical properties of this carbon precursor. In general, the glass transition temperature and aromaticity increase with decreasing volatile content. The effects of chemical composition and particulate inclusions on rheological properties are discussed.

The physico-chemical changes that take place when pitch is pyrolysed to carbon are described and interpreted via the construction of Transformation Diagrams.

The origins and control of porosity in manufactured carbon/graphite materials are discussed. It is shown that porosity is controlled largely by the degree of compaction of the coke grain (determined by the particle size distribution) and the volume fraction and carbon yield of the pitch binder.

The article concludes with a discussion of the surface activity of pitch and how it controls the interaction of the pitch with the coke grain during manufacture.

PHYSICAL PROPERTIES OF PITCH RELEVANT TO THE FABRICATION OF CARBON MATERIALS

B. Rand, A.J. Hosty and S. West

Dept. of Ceramics, Glasses and Polymers, Elmfield, Northumberland Road, Sheffield, S10 2TZ, U.K.

1 INTRODUCTION

Pitch is an essential precursor in the manufacture of a wide variety of granular engineering carbon and graphite materials, such as large scale graphite electrodes for the electric arc furnace, anodes and cathodes for aluminium smelting, nuclear graphite, electrical brushes and carbons for mechanical and wear applications. These materials are traditionally manufactured in a process outlined in Figure 1, in which a coke filler material is combined with pitch as a binder (Blackman 1970; Riley 1965). The pitch, on carbonization, generates a second carbon phase to bond what essentially becomes a particulate composite material. The coke fillers are derived from the carbonization of petroleum pitches and sometimes coal-tar pitch (delayed coking). During the pyrolytic conversion of pitch to carbon, there is an evolution of volatile matter which substantially reduces the volume yield of the secondary carbon phase and causes increased porosity in the product. Lower porosity products with increased strength and electrical conductivity are produced by impregnating the carbon stock with pitch, followed by recycling through the carbonization process. This stage may be repeated as many times as desired but the number of cycles is normally limited by economic considerations.

In recent years, pitch has increasingly been used in the fabrication of new high performance materials. These include carbon-carbon composites and carbon fibres with a variety of properties, e.g. general purpose, low cost fibres derived from isotropic pitch and high performance fibres from mesophase pitch. Figures 2 and 3 summarise the general fabrication processes for these materials (Fitzer 1987; Rand 1985). The pitch precursors used for the matrices in carbon-carbon composites and for granular products need to have similar characteristics.

The continued development of carbon products requires a sound scientific understanding of the physical and chemical properties of the pitch-like precursors used in their production processes. This Chapter focuses on the main physico-chemical properties of pitch relevant to these fabrication processes.

2 ORIGINS AND COMPOSITION OF PITCH
2.1 Coal-tar pitch

Coal-tar is produced as a by-product of the coking of bituminous coals to produce cokes; high temperatures (900°-1100°C) yield metallurgical coke and low temperatures (600°C) are used to produce domestic smokeless fuel. A small amount of tar is obtained in the low temperature processes. Pitch is produced from the coal-tar by distillation and heat treatment processes and Table 1 outlines typical components in different coal-tar fractions.

Table 1

Boiling Temperature /°C	Typical Components
350	Chrysene, Fluoranthene, Pyrene Phenanthrene, Anthracene, Carbazole
300	Acenaphthene, Fluorene, Methyl-naphthalene
250	Tar acids, Tar bases, Naphthalene
200	Benzene, Toluene, Xylene, Solvent-naphtha, Tar acids & bases

Pitch is the residue following the removal of the heavy oil (creosote oil), or anthracene oil, fraction. Typical distillation data are in Table 2.

Pitches are complex mixtures of many individual organic compounds; the precise composition and properties vary according to the source tar and the method of removal of the low molecular weight species. According to Smith et al. (1966), about two thirds of the compounds that have been isolated from coal-tar pitch are aromatic, the remainder being heterocyclic. Many compounds are substituted, the methyl group being the most prevalent. These isolated compounds contain from 3 to 6 rings and boil in the range 340°-550°C. Coal-tar pitch thus consists predominantly of the elements carbon and hydrogen with small amounts of nitrogen, oxygen and sulphur. The C/H ratio is an approximate indication of the degree of aromaticity of the pitch.

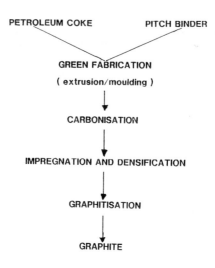

Figure 1. Flow diagram for traditional route to granular carbons / graphites.

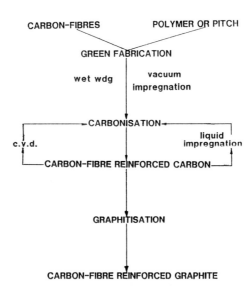

Figure 2. Flow diagram for production of carbon fibre reinforced carbon.

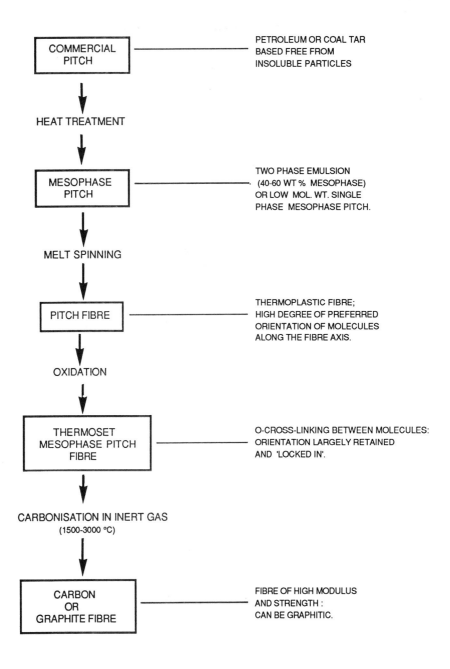

Figure 3 Flow diagram for the production of fibres from mesophase pitch.

Table 2

Products	Boiling range /°C	Weight %
Light oil	<195	1
Naphthalene oil	195-230	12
Creosote oil	230-300	6
Anthracene oil	>300	20
Pitch	residue	60
Tar acids		1

2.2 Petroleum pitch

Petroleum pitch is readily available. It is often obtained from catalytic cracker bottoms, i.e. the heavy residue from a catalytic cracking process, from steam cracker tar, a by-product of steam cracking of naphtha or gas oils to produce ethylene, or from any residues from crude oil distillation or refining. Pitch can be produced from these feedstocks by many different processes including thermal treatment, vacuum or steam stripping, oxidation, simple distillation or a combination of these as discussed by Newman (1976). As with coal-tar pitch, the chemical and physical characteristics are very dependent on the process and the conditions used, especially the process temperature and heat treatment time. Longer times and higher temperatures result in more aromatic pitches with higher anisotropic contents and, in general, higher "*glass transition temperatures*" (see section entitled 'Pitch as a Glassy Solid'). In general, petroleum pitches are less aromatic than coal-tar pitch. Table 3 summarises typical properties of binder pitches from the two sources.

Table 3

Characteristics of typical binder pitches from three different feedstocks

	SCT	Petroleum 1	Petroleum 2	Coal-tar 1	Coal-tar 2
Softening point (R & B/°C)	110	117	110	101	113
Coking value (wt% at 550°C)	52	54	56	57	60
Aromatic C/atomic %	78	82	80	89	88
C/H atomic ratio	1.37	1.44	1.57	1.77	1.76

(SCT refers to steam cracker tar)

2.3 Solubility as a characterisation technique

Solvent fractionation has widely been used as a method of characterization and, in recent developments, it has been effectively used as a method of modifying the chemical structure of pitch prior to and during the processing of fibres. Unfortunately, the coal-tar and petroleum industries have tended to use different solvents and nomenclature. The petroleum industry has defined the following species:

Carboids: insoluble in CS_2.

Carbenes: soluble in CS_2 but insoluble in CCl_4, (In general only minor quantities of these components are present).

Asphaltenes: insoluble in light paraffinic hydrocarbons such as n-pentane but usually soluble in CS_2, CCl_4 and C_6H_6. These are condensed, highly aromatic molecules.

Pre-asphaltenes: a term sometimes used to describe the fraction insoluble in solvents such as benzene but soluble in pyridine.

Pitches, particularly coal-tar pitches, used in the carbonization industries are often fractionated using the following solvents, although there is some variation in the nomenclature used by different workers as discussed by McNeil (1966).

Quinoline or pyridine: insoluble components are very high molecular weight aromatic compounds and solid impurities ("C_1" component or α– resins).

Benzene or toluene: insoluble material which is soluble in quinoline or pyridine is the 'β-resin' ("C_2") component, *i.e.* pre-asphaltenes.

Petroleum ether or n-hexane: Insoluble material which is soluble in benzene (toluene) is sometimes known as the "resinoid" fraction, identified with asphaltenes, whilst the soluble material is the "crystalloid" fraction.

Mack (1966) and later Riggs and Diefendorf (1979, 1980a, 1980b) discussed the solubility of pitches and aromatic compounds in terms of the theory of solubility of non-electrolytes (Hildebrand and Scott 1950). Solubility is determined by the homologous temperature, T/T_m, which should preferably be greater than 0.8 (T = system temperature, T_m = melting point of solvent), together with the cohesive energy densities of the solute and solvent, which determine their solubility parameters, δ:

$$\delta = \text{(cohesive energy density)}^{1/2} = (\Delta H_v - RT)^{1/2}/V_m^{1/2}$$

where ΔH_v = enthalpy of vaporisation,
V_m = molar volume.

At high homologous temperatures, the solubility as a function of solubility parameter passes through a maximum when the solubility parameters of solvent and solute are identical (Riggs and Diefendorf 1979). The solubility parameters of solvents normally used to characterise pitch are listed in Table 4. These solvents remove from the pitch those molecules of similar chemical characteristics. Some molecules may have a finite solubility in a range of solvents.

Table 4

Solvent	Solubility parameter $(Jm^{-3})^{1/2}$
n-pentane	14.45
n-hexane	14.96
carbon tetrachloride	17.63
toluene	19.53
benzene	18.75
tetrahydrofuran	20.29
carbon disulphide	20.49
pyridine	21.93

The solubility spectrum for a pitch can be useful in its characterization. Many of the molecules that are present in typical pitches are mesogenic in character and Riggs and Diefendorf (1979,1980a, 1980b) showed how they can be extracted and used to make instant mesophases often suitable for fibre drawing. The proportion and molecular weight distribution of mesogenic molecules in any particular pitch sample will vary with previous history.

2.4 Chemical characteristics

Pitches are complex mixtures of many individual organic compounds. The H/C ratio is a good indication of the average degree of aromaticity of the pitch but more sophisticated methods of characterization are now available. Solvent fractions can be analysed by vapour phase osmometry, for average molecular weight, and an indication of the molecular weight distribution can be obtained from gel permeation, high pressure liquid and gas chromatography (Bartle et al. 1979, Karr 1978 and Greinke and O'Connor 1980). The aromatic and aliphatic character of fractions can be determined by proton and ^{13}C nmr spectroscopy (Karr 1978, Fischer et al 1978, Ladner and Snape 1978).

2.5 Mesogenic character of pitch

Heat treatment of pitches converts them to the discotic, nematic liquid crystal state, known as the carbonaceous mesophase. The changes in volume fraction of this new phase as a function of heat treatment are well documented elsewhere, as are the different microstructural characteristics. This development can be understood as a lyotropic transformation in which the concentration of mesogens is raised above some critical concentration by the removal of non-mesogenic, or disordering, species in the volatilized matter. Riggs and Diefendorf (1979, 1980a, 1980b) proposed a schematic pseudo-Phase Diagram to show the effects of non-mesogen removal on the mesophase development. Because the mesogens and non-mesogens have different solubility parameters, they can be separated by solvent fractionation, enabling mesophase pitches to be produced without heat treatment. This results in materials with a more narrow distribution of molecular weight than in the thermal route because, in the latter, cracking reactions and molecular growth by combination of reactive fragments increase the size of the polynuclear hydrocarbon species in the melt. The composition of a pitch or mesophase pitch thus depends on how it is produced and, in the thermal route, factors such as the heating rate, time at some isothermal temperature and the nature of the gaseous atmosphere (e.g. whether or not hydrogenation can take place). The prevailing pressure and stirring/gas sparging conditions are all of major significance as the balance between loss of molecules by straight evaporation and as products of the cracking reactions is changed. All these factors affect the physical properties.

3 STRUCTURE OF PITCH
3.1 Pitch as a glassy solid

Isotropic pitch falls into the category of glass-forming materials, one important characteristic feature of which is that there is no defined melting point, but the material passes through a temperature region, the "glass transition region", before forming a viscous liquid. The nature of the glass transition is of fundamental importance to the properties of all glass-forming systems. At the glass transition, there is no discontinuous change in the first derivatives of the molar free energy, but the second derivatives do show a discontinuity, e.g. in the heat capacity, the thermal expansion coefficient and in the isothermal compressibility. In the transformation from the crystalline state to the isotropic liquid, there is a loss of positional order and a gain in molecular or conformational mobility, but in the glass transition there is only a gain in mobility and, in the case of the rigid molecules in pitch, the conformational effects are minimal. The entropy change at the glass transition reflects only this change in molecular mobility.

The liquid crystal phase which develops from pitch systems differs from those of other small molecular systems in that, on cooling, a truly crystalline phase is not produced. This is clearly demonstrated by the X-ray diffraction patterns of mesophase-pitch systems at ambient temperature. In this respect, they resemble certain liquid crystalline polymers in that they solidify without change in positional order, i.e. they pass through a glass transition temperature. A more precise name might be glasses with orientational order, but the term mesophase glass will be used here to indicate the origin of the phase and to distinguish it from the original pitch system.

The glass transition is perhaps best considered as a kinetic phenomenon. As the temperature is lowered, the relaxation times for molecular translations and rotations are drastically increased until in the glass transition they become so long that equilibrium structures cannot be developed. This means that the structure developed at the glass transition, and the value of Tg measured, is dependent upon the cooling rate. These effects for glasses are extensively reviewed elsewhere. However, the fundamental significance of the glass transition is that it is the temperature at which relaxation processes become important in realistic time scales. This determines the temperature range over which those properties that depend on these relaxation processes change rapidly and thus the temperature dependence of properties must be related to the glass transition temperature or some other characteristic temperature dependent upon Tg. Thus, the glass transition temperature becomes the key factor in determining ALMOST ALL OF THE PHYSICAL PROPERTIES of a molecular mixture such as pitch.

Free volume

The glass transition is usually discussed in terms of the fractional free volume in the system. There are various definitions of free volume but the essential feature is that the glassy state is considered to be one of constant free volume, i.e. that volume not occupied by the molecules themselves. Above Tg, free volume changes rapidly with temperature, accounting for the change in relaxation times and of thermal expansion coefficient. The significance of this in determining properties is discussed in the section entitled 'Rheological Properties of Pitch'.

3.2 Pitch as a colloidal system
The isotropic pitch matrix

Many pitches are undoubtedly colloidal systems in the sense that they contain dispersed, finely-divided particulate matter. However, there is also a school of thought that the optically isotropic pitch matrix is itself a micellar colloid. Early considerations are reviewed by Mack (1966). These concepts have been adapted by Riggs and Diefendorf (1980a) who suggest that an intrinsic pitch solution is unfavourable on thermodynamic grounds due to the excessive heat of mixing between the large aromatic molecules and the paraffinic and naphthenic components. They rationalised the apparent homogeneity by suggesting a micellar system in which the aromatic species 'sit' in the centre of the micelle, surrounded and 'solubilised' by the naphthenics, which, in turn, are surrounded by the paraffins. In this model, the central portion of the micelle is occupied by the mesogens and the concept of Riggs and Diefendorf (1979, 1980a, 1980b) is that once this delicate balance is disrupted, the mesogens can separate as an additional phase. If this model is applicable, then the physical properties of the pitch systems should change dramatically as the micellar arrangement is changed, for example, by temperature. The existence of randomly distributed basic structural units of aromatic species in the form of stacks of the order of 1.0 nm in size has been detected in various asphalt type precursors by Monthioux et al. (1982) using high resolution electron microscopy. Exactly how these domains relate to the micelles in Diefendorf's model remains problematic at the present time.

3.3 Particulate inclusions

Many pitch materials are clearly dispersions containing one or more types of dispersed particulate phases. Basically, these are of two types, primary and secondary. The primary particles are present in material from which the pitch is produced. This type is most common in coal-tar pitches and arises mainly from the coke ovens. The particulate matter may be inorganic, deriving from the refractories, or carbonaceous, arising from the gas phase pyrolysis of vapours. The particles so produced, however, are not identical to carbon black, having, in the main, larger particle size. The secondary type of particle is produced by heat treatment of the pitch and may occur in all types of pitch subjected to this type of treatment. The phase is "mesophase" (see Chapter 2) and its size distribution may vary quite widely. The effects of the two types of inclusions on the physical properties of the pitch can be expected to be quite different, especially at elevated temperatures when the mesophase inclusions will, at temperatures greater than T_g, have a measure of fluidity. Detailed studies of particulate inclusions are described by Marsh et al. (1985), Tillmans et al. (1978) and Kremer (1982). The particulate matter is often referred to as QI, quinoline insoluble matter. However, whilst all primary particulates are insoluble in quinoline, the mesophase may be partly soluble and there may be species in the isotropic matrix that are insoluble in quinoline, so this is not an absolute measure. Nevertheless, the QI fraction is a very important factor in determining the properties of pitches, particularly binder pitches. For example, the coke yield increases with increasing QI. The effects of these particulate inclusions are discussed in later sections. It is clear from the above that a pitch may be:-

i) an apparently single phase, but perhaps is a colloidal dispersion of fine micelles,
ii) a two phase material in which the minor phase is dispersed solid matter (primary QI),

iii) a two phase material comprising an isotropic and an anisotropic phase the proportions of which can vary,
iv) a single anisotropic phase, i.e. all mesophase (bulk mesophase),
v) multiphase, in which dispersed solid matter is present in addition to ii) and iv) above.

These different compositions inevitably lead to marked differences in properties, particularly the rheological properties.

4 RHEOLOGICAL PROPERTIES OF PITCH

A temperature region of great importance in the fabrication of carbon/graphite artefacts is that region where the mixing and wetting of coke granules or carbon fibres takes place and here the flow properties of the pitch are of great significance. These flow properties are very strongly affected by temperature and this aspect is dealt with in some detail. It is essential that the viscosity be low enough to allow flow into capillaries in infiltration stages and the viscosity of the pitch is critical in determining the extrusion and moulding characteristics of pitch-coke mixtures.

4.1 Newtonian and non-Newtonian flow

The flow characteristics of apparently single phase pitch materials at temperatures of the order of 100°C above T_g are normally Newtonian in character, i.e. the variation of rate of shear strain (shear rate, 'γ') with the shear stress, 'τ', is linear, such that the viscosity coefficient, $\eta = d\tau/\gamma g$, is independent of shear stress, as shown in Figure 4. This basic type of behaviour may be altered if there are changes in the molecular interactions or in the structure of the fluid as the flow process takes place, i.e. with changing shear rate. For example, a change in the extent of preferred orientation of molecules during shear can alter the viscosity coefficient quite markedly. Such effects are common with liquid crystal systems (see Chapter 2) and can therefore be expected for mesophase pitches. The inclusion of particulate matter in a suspension often leads to attractive interactions between the particles which agglomerate to form flocs, the progressive disruption of which under shear can also markedly affect the form of the flow curve, leading to the presence of apparent yield stresses and a viscosity coefficient which decreases with increasing shear rate, Figure 4. In this case, the term apparent viscosity, defined as τ/γ is often used. But, strictly, the rate of shear at which the measurement is made should be specified. It is clear that for non-Newtonian systems, shown in Figure 4, more than one parameter is required to describe the flow process.

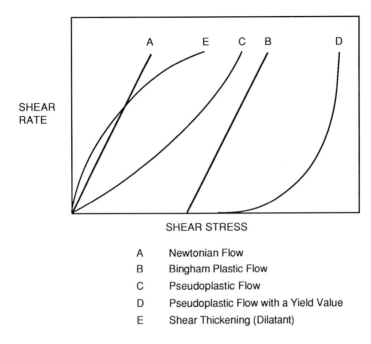

Figure 4. Non-Newtonian flow behaviour of various types.

A   Newtonian Flow
B   Bingham Plastic Flow
C   Pseudoplastic Flow
D   Pseudoplastic Flow with a Yield Value
E   Shear Thickening (Dilatant)

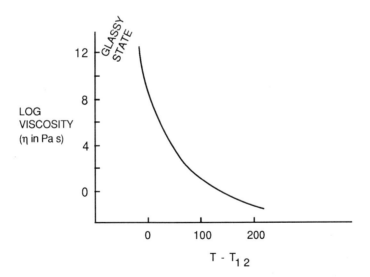

Figure 5. Typical viscosity-temperature curve for an organic glass-forming liquid above the glass transition temperature (taken as the temperature at which $\eta = 10^{12}$ Pa s, T_{12}).

4.2 Effect of temperature on the viscosity coefficient

The importance of T_g as a characteristic parameter is illustrated by reference to Figure 5, which shows the variation of viscosity 'η' with temperature for a typical glass forming system. It is assumed that the region of viscosity above 10^{12}Pa s corresponds to the glassy region, an assumption that is often made. The viscosity is seen to be dependent upon $T-T_g$, and the significance of T_g is seen in its control of the temperature region where specific viscosity values are obtained. It is often desirable to have mathematical expressions describing the temperature dependence of viscosity and it seems obvious that the most rigorous expressions will incorporate a reference temperature that will in some way account for the increase in free volume with temperature. Such equations are widely, and relatively successfully, employed in describing the η-T relationships for polymers (Ferry 1969) and for inorganic glasses. Also, they have been used in representations of the temperature dependence of visco-elastic parameters for asphalts used in road building. This aspect of pitch rheology has been reviewed recently by Rand (1987) who emphasised that the most successful and the most useful expressions include a reference temperature (i.e. some suitable isoviscous temperature) which is, or can be related to, the glass transition temperature. The determination of suitable isoviscous reference temperatures is an important aspect of the characterisation of pitch precursors for carbon fabrication. However, it is important to realise that the reference temperature, whilst being important in locating the temperature range for particular viscosity values, is not the exclusive parameter determining the precise form of the viscosity-temperature curve for the pitch. It is established that pitch materials with the same reference temperature show somewhat different viscosity-temperature relationships.

4.3 Measurement of the glass transition temperature

The following are the most widely used techniques for determination of T_g:-
  1) Volume dilatometry,
  2) DSC (change in C_p),
  3) Change in visco-elastic properties,
  4) NMR, ENDOR techniques.

Although the most direct technique is the change in thermal expansion coefficient, the most widely used method is probably DSC. Rand and Shepherd (1980) found that this technique was difficult to apply to heat treated pitches and this was confirmed by Barr and Lewis (1982) for mesophase pitches. In these systems, the transition is broadened extensively and the change in baseline, characteristic of the transition, was difficult to observe. Spectroscopic methods, such as broad-line proton nmr or ENDOR techniques, are excellent for determining T_g values. The spin-lattice relaxation times in the nmr technique and the electron-nuclear dipolar interactions in the ENDOR case are very sensitive to molecular motion. The techniques have been applied to polymer systems and to pitches. Dynamic mechanical thermal analysis has not been extensively used for pitches to date.

Other reference temperatures

It is common practice to characterise the rheology of pitch materials, not by measurement of Tg, but by determination of so-called softening points. These parameters are estimated by various empirical standard methods which are described by Traxler (1964), King (1978) and Abson and Burton (1964). Essentially, they determine isoviscous temperatures, although not with the precision allowed by more precisely defined technqiues. For example, the Ring and Ball method locates an isoviscous temperature where the viscosity is approximately 10^3 Pa s.

4.4 Factors determining the glass transition temperature and other reference temperatures

The relationship between the glass transition temperature and other isoviscous temperatures has been discussed by Rand (1978). Obviously isoviscous temperatures increase and decrease as Tg is determined by the molecular composition of the pitch system. Unfortunately, there have not been many detailed studies of these relationships. An insight into the most important factors can be obtained from the literature on polymer systems where it is found that, in general, Tg increases with:-

1) increasing average molecular weight, M_n, (for relatively low molecular weight polymers, i.e. comparable with pitch molecules,
    $$T_g = T_{g\infty} - (K/M_n)$$
    where $T_{g\infty}$ is the value of T_g at infinite molecular weight),
2) increasing inter-molecular forces, i.e. molecular structure,
3) decreasing plasticizer concentration; the effect, according to Fox (1956) is described by:-
    $$1/T_g = w_1/T_{g,1} + w_2/T_{g,2}$$
    (w_1 and w_2 are the weight fractions of polymer and plasticizer)

Thus, the important factor for pitch systems should be the molecular weight distribution in which the smaller molecules, the disordering species in Riggs and Diefendorf's model (1979, 1980a 1980b), might be expected to act as a plasticizing component and have a disproportionate effect on Tg. For pitch systems that have not been deliberately treated with long chain polymers and especially those that have received substantial heat treatment, the larger molecules are also likely to be more rigid, planar species with strong intermolecular forces. There should, therefore, be a correlation between the value of Tg and the spectrum of solubility parameters shown by the pitch. The variation of Tg with toluene insoluble content (TI) is shown in Figure 6. The volatile content, TI and the H/C ratio are also inter-related.

Hayes (1961) has shown that Tg can be linearly related to the cohesive energy density for a series of polymers. When such a relationship is applied to pitch systems, then Tg of solvent fractions is linearly related to the solubility parameter, d_2. Barr and Lewis (1982) expressed Tg data in terms of the Fox equation for blends of pyridine-soluble and insoluble-fractions, assuming that the soluble fraction would act as a plasticizing species. A linear relationship was obtained. A similar relationship was shown for blends of acetone-soluble and insoluble-fractions from an ethylene cracker tar (Rand 1987). Wallouch et al. (1977) combined pitches in different proportions. They found that the Ring and Ball softening point of the mixture could be expressed in terms of the values for the original pitches by a simple mixture rule expression. These studies all show that the value of Tg and other isoviscous temperatures are determined by the pitch composition. However, the precise relationships are not yet fully established and more detailed work is required before a satisfactory model can be developed.

There are very few measurements reported of Tg values for single phase mesophase pitches. Riggs and Diefendorf (1983) show that, for mesophase pitches produced by solvent fractionation, there is a linear relationship between Tg and the solubility parameter of the solvent used. A linear relationship also applies to Tg vs cohesive energy density, consistent with Hayes' work (1961) discussed above. Benn et al. (1985) showed that the Tg of the mesophase spheres produced during pyrolysis of an ethylene cracker tar pitch was fairly constant (approximately 150°C) until the phase inversion point was reached, after which it increased rapidly with further decrease in the volatile content.

Barr and Lewis (1982) report Tg values for mesophase pitches but it is not clear whether they also contained isotropic phase. Many of the values reported by Whitehouse and Rand (1982) refer to mesophase pitches that are two phase systems. For such systems, each glassy phase will have its own Tg value. Because mesophase and isotropic phase are of quite different molecular character and molecular weight, Tg values might be substantially different, with the mesophase value being much the larger of the two. Thus, penetration methods of determining Tg probably reflect the Tg of the continuous phase. One of the reasons why Tg is difficult to determine for heat-treated pitches is that, being of two phase character, the change in the characteristic property being measured is reduced due to the dilution effect of the second phase.

4.5 Effect of particulate matter on rheology

The presence of a dispersed phase raises the viscosity at all rates of shear and may introduce non-Newtonian character to the flow behaviour in shear flow. The effect, however, is dependent on many factors including the volume fraction, particle size distribution and strength of particle-particle interactions. Particularly important is whether or not the second phase is solid or an immiscible liquid and, if the latter, then its rheological behaviour becomes of significance. However, it is clear that if this second phase is a glass-forming system, such as the carbonaceous mesophase, then whether or not it behaves as a liquid depends on its Tg value. The two general types of particulate matter can, therefore, be expected to show different η-T behaviour. Of course, as is pointed out above, the precise behaviour will depend on several factors. Knowledge of the Tg value of secondary particulates could therefore be useful. It is clear that the presence of such dispersed phases will render a mathematical description of η-T data difficult.

4.6 Mesophase rheology

Little is really understood about the detailed flow behaviour of two phase mesophase pitches. In-situ studies have demonstrated the non-Newtonian character of flow behaviour but, because these have often been determined in reactive conditions, it is not certain that they are accurate representations of the true flow behaviour of a particular system at a specified extent of transformation to mesophase. Evidence of thixotropic behaviour has been reported. In the experiments reported by Collett and Rand (1978), this effect was shown under non-reactive conditions but it was also found that the mesophase was almost solid-like. The detailed nature of the thixotropic character of fluid mesophases, if present, remains to be properly characterized. Although shear thinning effects have been reported for two phase mesophase pitches, it is not yet clear to what extent the effect is due to shear orientation of the anisotropic phase or to shear induced rupture of coagulated/coalesced mesophase, nor has the reversibility been studied in detail. Perhaps the most rigorous investigation of the flow behaviour of mesophase pitch has been presented by Nazem (1982). He studied mesophase pitches which were entirely or predominantly mesophase. Shear thinning behaviour was observed in some cases but in others, the behaviour was apparently Newtonian. These 'Newtonian' systems, however, could also show evidence of viscoelasticity (indicated by the Weissenberg effect and by die swell measurements).

A study of the extensional flow properties of well characterized mesophases is also essential for an understanding of the spinning characteristics of mesophase pitches in fibre production. Such a study has not yet been undertaken, probably because our understanding of the development of the carbonaceous mesophase and how to control its properties is now only just emerging. In the use of mesophase pitches, either for fundamental studies, or as a precursor material in the manufacture of engineering carbons and graphites, characterisation of the precursor pitch in terms of its composition is essential. This is to establish the thermal reactivity of the material and, in particular, to determine the temperature regime in which, on heat treatment, the pitch begins to decompose, changing its compositional characteristics and associated physical (rheological) properties. Such information is required in sophisticated processing operations utilising such precursors.

5 PYROLYSIS OF PITCH

The conversion of pitch to carbon takes place by a process of pyrolysis. There are several reviews of this process and of the significance of the intermediate mesophase. The essential processes that take place are:-
   a) <u>Evaporation of low molecular weight species</u>
   As the temperature is progressively raised, species of increasing molecular weight are volatilised. The process is controlled by:-
   i) the surface evaporation rate as determined by the temperature, pressure, gas phase and liquid phase composition, and
   ii) the kinetics of diffusion of the volatile species to the surface.

Thus, for an unfired pitch-bonded coke system, the weight loss as a function of temperature depends upon its dimensions as well as those factors listed above. This is discussed in some detail by Fitzer <u>et al</u>. (1971). In the production of mesophase precursors for advanced carbons such as fibres, it is common practice to use gas sparging and vigorous stirring conditions to facilitate the removal of the volatile species.
   b) <u>Cracking reactions followed by evaporation of volatile fragments</u>
   This takes place at higher temperatures, of the order of 350°-400°C. The fragments mainly result from the thermal scission of aliphatic side-chains to polycondensed aromatic ring structures. The polynuclear aromatic radicals produced are quite reactive and combine to produce planar aromatic ring structures of even higher molecular weight and greater aspect ratio. These remain in the melt altering its physical characteristics quite markedly. When this stage is operative, the average molecular weight of the volatile species decreases as there is a change in the proportion of volatiles emanating from unreacted volatile molecules and low molecular weight products of the cracking reactions.

An important criterion in selecting a pitch precursor for carbon manufacture is the carbon yield. This depends largely on the composition of the precursor pitch but also it is strongly influenced by the pyrolysis conditions. Increasing the heating rate, the pressure and the 'volume to surface ratio' all tend to increase the carbon yield by restricting the evolution of volatile molecules present in the original pitch. The retention of these species in the carbonizing liquid as the temperature is raised allows them to participate in cracking reactions and in molecular growth processes. For a fixed set of pyrolysis conditions, the carbon yield is influenced very strongly by the proportion of asphaltenes, pre-asphaltenes and insoluble carbonaceous particulates present in the original pitch. As an illustration, Figure 7 shows the variation of volatile content with toluene-insolubles material. Volatile content, here, is defined as the percentage weight loss of a pitch precursor on carbonization to 800°C at a fixed heating rate, pressure and flow rate of inert gas and is, thus, related to the carbon yield.

5.1 Transformation diagrams

A useful way of representing the pitch-mesophase-carbon transformation is the Transformation diagram of Whitehouse and Rand (1985). This is shown schematically in Figure 8. Essentially, some average compositional characteristic such as H/C ratio or volatile content is used as a measure of the extent of transformation of the material towards carbon and this is related to characteristic temperatures such as the glass transition temperatures of the phases present. In this way, the changes in the structure and properties of the pyrolysing residue can be related to the extent of volatilisation. A precursor used in the fabrication of a carbon artefact could be the pitch product at any stage of such a pyrolysis process. On reheating this residue, at a temperature determined by the heating rate, prevailing gas pressure and composition, it begins to decompose again and therefore included in the diagram are the thermal decomposition temperatures of the materials. (NB. This is the decomposition temperature of the total material not of each individual phase). In this way, a measure of the thermal stability of the precursor is displayed and related to parameters that reflect the composition and temperature dependence of the physical properties of the precursor. Of course, only a limited amount of information can be displayed on the diagram. Furthermore, the diagram can be used to follow the changes in the properties of the pitch precursor that take place during pyrolysis. It is important to note, as outlined above, that different pyrolysis conditions result in quite different distributions of molecular weight and type at the same value of some average parameter. Therefore, the form of the transformation diagram for any precursor pitch is not unique but is dependent upon pyrolysis conditions. However, the diagram itself then becomes a useful way of presenting data to illustrate the effects of different pyrolysis conditions. Also, it is suggested that the diagram is useful in focusing attention on the fundamental nature of the pitch-mesophase-carbon transformation itself.

Figure 8 shows that the volatile content and H/C atomic ratio are high when the pitch is entirely isotropic, reflecting the matrix phase of a typical binder pitch or

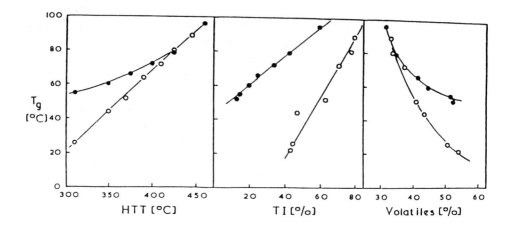

Figure 6. Variation of glass transition temperature with toluene insoluble content and with volatile content for pyrolysis residues from Ashland A240 (●) and a coal extract pitch (o).

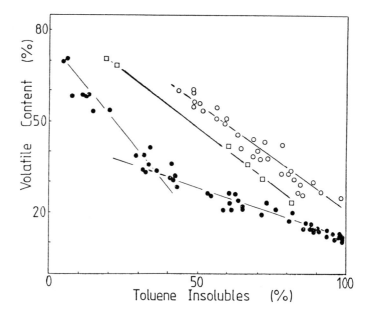

Figure 7. Variation of volatile content with toluene insoluble content for pyrolysis residues from A240 pitch (●), and two coal extract pitches (o, □).

impregnation pitch. On heating beyond the glass transition temperature, the viscosity decreases rapidly and at T_0 volatilization sets in and the composition begins to change, the rate of change depending upon the temperature, atmosphere, stirring, etc. This brings about a change in the Tg and the viscosity-temperature curve for the material is shifted to higher temperatures. When mesophase appears (or is already in the precursor pitch because of previous processing), there are two glass transitions to consider, that of the mesophase being the greater of the two. If, as might be anticipated, the composition of both phases continue to change as pyrolysis proceeds, then the Tg of each phase should progressively increase with decreasing volatile content, H/C, etc. Thus, the Tg values in the two-phase region establish the compositional and temperature regions in which there exist two-phase mesophase pitches which behave as emulsions, or as suspensions of 'solid-like' mesophase particles/agglomerates or as two-phase glasses. In this two-phase region of the diagram, the decomposition temperature shown is that of the system as a whole which is determined by the most reactive phase, almost certainly the isotropic phase. The decomposition temperature of the mesophase is probably much higher, reflecting the higher molecular weight. The lower decomposition temperature, however, determines the thermal stability of the total system.

At low H/C atomic ratios and volatile contents, only one phase remains, the mesophase, and it is in this region that pronounced changes take place both in composition and in associated rheological properties. Hence, the Tg increases very rapidly with decreasing volatile content. This is an important region of the transformation to carbon because it is here that the microstructure of the ultimate carbon product is finally determined, when the relaxation processes in the material become slow as its Tg approaches that of the reaction temperature and the microstructure prevailing at that stage is 'locked-up'. In the diagram, Tg is seen to correspond to the decomposition temperature at a certain point. For the particular pyrolysis conditions pertaining to the diagram, this point establishes the upper limit at which a stable mesophase pitch can exist in the liquid crystalline state. Systems with slightly lower volatile contents still show a glass transition and pass into a liquid crystalline state but the composition of the material when in that state changes continuously with time and temperature.

It has been shown above how the viscosity of a pitch/mesophase pitch relates to its Tg. Thus, the value of T-Tg above Tg largely determines the viscosity at any particular rate of shear and the proximity of the decomposition temperature to the Tg will determine the minimum viscosity attainable in a stable pitch/mesophase pitch system at any average composition. In principle, the viscosity-temperature relationships, in the stable region, could be included in the diagram as isoviscous temperature lines, as shown by Whitehouse and Rand (1987).

5.2 Uses of the transformation diagram

In the spinning of mesophase-pitch fibres, a continuous 100% mesophase pitch is desirable. The position of such a material in its transformation diagram shows the viscosity range that can be attained before thermal degradation takes place. Similarly, when the spun fibres are carbonized, molecular mobility above the glass transition temperature can lead to loss of preferred orientation. If T_o is close to T_g then this relaxation is small but dependent on the rate of heating above T_o. However, such a material would not be spinnable in a stable state. Consequently, it is necessary to use a pitch lower down the transformation range and to thermoset the spun fibre by an oxidative process prior to carbonization. Even here the value of the T_g is critical because if the thermosetting temperature is above T_g, some molecular relaxation may take place. Stevens and Diefendorf (1986) have commented on the effects of oxidation in the vicinity of T_g.

5.3 Experimental diagram

Figure 9 shows an experimental diagram for an ethylene cracker tar pitch. The form is more or less as predicted in the schematic diagram discussed above. One important observation is that, in the two phase region, the Tg of the mesophase does not change as predicted in the schematic diagram until the phase inversion region is approached, after which it changes very rapidly. This suggests that the mesophase composition does not change very greatly in the two phase region and the change in composition reflects mainly the increase in the relative amounts of mesophase and a progresive change in the composition of the isotropic phase towards that of the mesophase. This is concluded from the increase in the Tg of the isotropic phase, as shown in the diagram. Greinke (1986) and Greinke and Singer (1988) report similar conclusions in the region of the phase inversion on the basis of chromatographic investigations of "solubilized" mesophases.

6 PITCH AS A BINDER AND MATRIX MATERIAL IN ENGINEERING MATERIALS
6.1 Effect on porosity

As explained in the introduction, pitch is widely used as a binder and impregnant in the fabrication of granular composite graphites and, using infiltration processes, it provides a source of carbon matrix in carbon-carbon composites incorporating carbon fibres (CFRC and CFRP, see Chapter 6.). The function of the carbon phase is to fill, as efficiently as possible, the intergranular/interfibrillar space, to minimise the porosity which has a major influence on all the physical properties of the carbon/graphite product. The matrix microstructure and the interaction of the matrix with the granular/fibrillar components are critical in

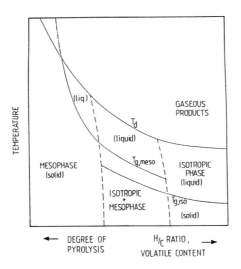

Figure 8. Schematic form of a transformation for the pitch-mesophase-carbon transformation.

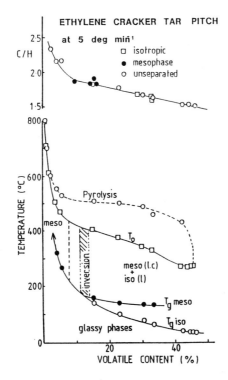

Figure 9. Experimental transformation diagram for an ethylene cracker tar pitch undergoing pyrolysis to carbon at a linear heating rate of 5 K min^{-1}.

determining thermo-mechanical properties, although this aspect is beyond the scope of this Chapter. The materials technologist is required to engineer a material with a controlled microstructure such that all major properties are optimized. This requires a fundamental understanding of how the starting materials behave during processing and this includes knowledge of the origins of the various types of pore that may result.

Pores may, in general, arise from the following sources:-

1. Pores within the filler phase (more important with coke fillers than with fibres).
2. Intergranular/interfibrillar pores not completely filled by pitch due to inefficient mixing/infiltration or use of insufficient pitch.
3. Pores generated within the binder/matrix phase during carbonization.
4. Interfacial pores, between binder and granular/fibrillar phases due to separation during processing.

Pores within coke grains are present in the starting material but further porosity develops if the granular material is heated during processing to a higher temperature than it has previously experienced. Intra-granular porosity (if it is open porosity) may be partly filled by the binder phase during the initial mixing operations. This is determined by the wetting and flow characteristics of the pitch which collectively control its capillarity. However, the degree of penetration may be modified during subsequent processing because, as the pitch composition changes, so will its surface activity and rheology.

Pores within the binder phase arise due to the substantial volume shrinkage during carbonization, when the weight change may be of the order of 50% and accompanied by a substantial increase in its true density. This results in a volumetric yield of approximately 33%. Hence, in the production of high performance composites, pressure pyrolysis is used to increase the yield to 85 wt% or so, giving approximately a 55% volumetric yield.

Interfacial pores depend on the adhesion between binder and filler phases and the extent to which this is maintained when the binder shrinks. If adhesion is poor, the binder phase may separate during carbonization, opening up interfacial pores. However, if adhesion is maintained, then the contraction of the matrix may 'pull' the grain/fibre and an overall shrinkage of the body (decrease in bulk volume) ensues. Whether this takes place or not will depend on the packing of the filler material. If the filler is efficiently packed, then movement of grains/fibres relative to each other is allowed only if it leads to a volume expansion. Thus, control of porosity is, at least partly, influenced by the interfacial interactions between the constituent phases. Another aspect is important for the filler phase. If the packing is efficient and the intergranular/interfibrillar space is filled with the binder phase prior to its carbonization, then it will be difficult for the volatile matter to escape, pressure of volatile matter may build up locally and create 'bubble-like' or so-called 'gas-entrapment' pores in the carbonization region where the binder has some fluidity but is of high viscosity. This can cause a volume expansion of the product (bloating), resulting in an increase in porosity over that due to contraction of the binder on carbonization.

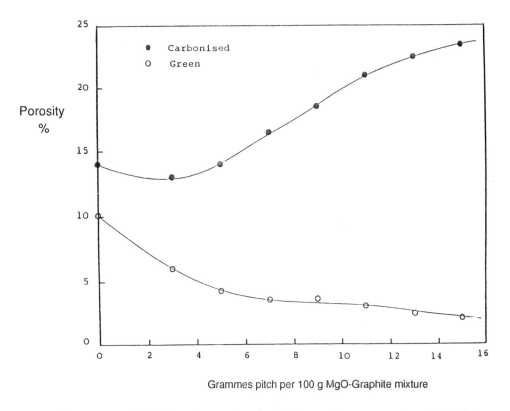

Figure 10. Variation of porosity of a MgO-graphite composite with initial content of pitch binder before (o) and after (●) carbonization of the binder at 1500°C.

The role of a coal-tar pitch binder in determining the porosity of a granular compact is illustrated in Figure 10. The example is actually a compacted composite of magnesia and natural flake graphite but the principles are apposite. The variation of porosity with binder content is shown in the green state. The porosity is reduced to very low levels indeed as the pitch binder fills the interstices of the compact. However, on carbonization of the binder, the porosity is quite substantially increased. Those binder contents that gave the lowest porosities, prior to firing, actually show the highest porosity values after carbonization. The minimum in porosity, here, is located at an initial pitch content of 3-4% by weight of the oxide-graphite mixture. At this point, the binder is almost filling the interparticulate space. When the pore structure is filled by binder phase, the release of volatile matter is restricted and bloating, leading to an increase in bulk volume of the composite, occurs. Furthermore, increasing the binder level above this point is deleterious in terms of porosity development because the volume shrinkage in that phase increases the porosity substantially. This is why such high porosities are obtained in the carbonized products above about 8 wt% binder. It is clear that this limit of binder content varies from product to product as the interparticulate volume varies due to different grain size distributions and/or particle shapes. The binder levels used in the manufacture of oxide-graphite composites are much smaller than those conventionally used in the fabrication of typical pitch-coke materials because the interparticulate volumes are lower in the former case due to the efficient packing of the flake graphite.

The optimum binder content for granular carbons/graphites varies according to the size distribution of the coke filler, being higher, the smaller the average particle size, reflecting the higher surface area of smaller particles and their tendency not to flow and pack efficiently. It has been shown that there is a maximum in the Young's modulus of polygranular bodies as the initial binder content is increased (Hutcheon 1970).

6.2 Surface activity of pitch

The importance of carbon yield and the filling of intergranular or interfibrillar space was emphasised previously. The surface activity of pitch is of the greatest significance in determining the latter and also controls the filler-matrix interactions. Pitch is required to wet and flow along carbon surfaces which, in different products, have various degrees of structural perfection. Indeed, in the fabrication of carbon-carbon composites by the liquid infiltration route, there may be different carbon surfaces present simultaneously, e.g. generated from resin, from pyrolytic carbon and from carbonized pitch as well as carbon fibres, which themselves can have different extents of graphitization. This surface activity is also intimately associated with the rheological properties of the pitch. The surface activity is determined by the surface tension of the pitch and by the physico-chemical interaction of the pitch with the carbon surface, which determines the pitch-carbon interfacial free energy.

Surface tension

There are only few measurements of the surface tension of pitch materials, due largely to the difficulties in carrying out this measurement. Some of these are discussed by Mack (1966). Greenhalgh and Moyse (1970) showed that the surface tension decreases approximately linearly with temperature over the range 130°-170°C for a pitch with a Ring and Ball softening point of 300°C. Results nearer to T_g could not be obtained. The addition of creosote oil lowered the surface tension, indicating the importance of pitch composition, but no detailed analyses of the relationships between composition and surface tension have been reported for pitches. Since the surface tension is a manifestation of intermolecular forces, it should be related to other properties which are determined by the same intermolecular forces, e.g. cohesive energy density. This remains to be established, however.

Contact angle and wetting studies

The wetting of a solid by a liquid is often characterised by the contact angle, θ, which is related by the Young equation to the solid-liquid, $γ_{SL}$, solid-vapour, $γ_{SV}$ and liquid-vapour, $γ_{LV}$, surface energies. The criterion for wetting is that the contact angle is less than 90°, i.e. $γ_{LV} < γ_{SV}$. Various studies have been made of the contact angles of pitch on cokes and other carbon surfaces. Often, granular beds of coke particles have been used and this can lead to difficulties in interpretation because the role of porosity must be taken into account. This was recognised by Lahaye et al. (1976) who showed that the contact angle is determined by the concentration of surface complexes on the carbon surface, up to a certain critical level. These complexes only form on the edges of graphitic layers and the proportion of a carbon surface that comprises edge surface can vary widely from carbon to carbon or with heat treatment temperature. Thus, it seems likely that the contact angle for a specific pitch at a specified temperature increases with heat treatment temperature of a carbon and from carbon type to carbon type. There is some evidence to support this concept.

Lahaye et al. (1976) and also Greenhalgh and Moyse (1970) showed that the contact angle decreases with increasing temperature at temperatures above the softening point. The effect varied with pitch source. Greenhalgh and Moyse (1970) showed that the addition of surface active agents and creosote oil decreased the contact angle and that the adsorption of vapours from the pitch on carbon surfaces had a strong influence on measured values. Pre-treatment of the carbon surface with pitch vapours and acetone brought about a decrease in contact angle also. These adsorption effects are responsible for a time dependent wetting behaviour and hysteresis in the measured contact angle.

In many situations it is the flow of pitch into capillaries that is of significance. Flow of a liquid into a small pore can take place under the influence of capillary action. For a cylindrical capillary of radius, r, the capillary pressure, P, is given by:-
$$P = 2 γ \cos θ / r$$

The temperature dependence of γ in studies of pitch is much smaller than that of θ so P will tend to increase with temperature and there will be a greater tendency for pitch to be drawn into small voids. The capillary pressure is greater for small capillaries than for large. This means that the equilibrium state is one where the fluid moves from the large to the small pores. In a dynamic infiltration situation, however, the large pores will fill first with fluid because the flow rate in these is larger, as shown by the Poiseuille equation:-

$$\mu = r \Delta P / 8\eta \, l$$

where μ is the linear flow rate, l is the wetted length of cylindrical pore and ΔP is the applied pressure. The viscosity of pitch changes quite markedly during carbonization, attaining values much lower than those normally prevalent at the extrusion stage for pitch-coke type materials. Thus, the pitch phase may penetrate hitherto inaccessible regions or even exude from the pores of the granular mass.

In assessing the suitability of pitches for impregnation purposes, a simple capillary flow test has often been adopted (Greenhalgh and Moyse 1970), in which a drop of pitch is placed on the surface of a granular mass of coke and heated at a linear rate. The change in dimensions as the pitch wets and flows into the porous medium is followed as a function of temperature. This test is able to rank pitches in terms of their wettability and infiltration behaviour but detailed interpretation of the data is difficult, although attempts have been made (Ehrburger and Lahaye 1984). One factor which is of great significance in the selection of pitches for impregnation purposes is the presence of particulates (QI). As outlined earlier, the behaviour of secondary particulates might be expected to be quite different from that of the primary type, particularly if the temperature is significantly above the T_g of the secondary phase, when it can coalesce and effectively seal the surface of the article to be impregnated.

7 ELECTRICAL CONDUCTIVITY

Few definitive studies have been made into the subject of the electrical conductivity of pitch even though 70% of the annual world pitch production is used in the manufacture of electrodes (Kremer 1982). At temperatures below about 100°C, the resistivity is of the order of 10^{10} Ωm but, after heat treatment to 400°C or so, it can attain values of the order of 10^2 Ωm, typical of silicon or germanium at room temperature.

The nature of the conduction processes in pitch is as yet unknown, but several possibilities have been proposed including charge transfer complexes between pitch molecules and the inorganic impurities therein (Speight and Penzes 1978; Sharp 1987), viscosity controlled carrier diffusion and a solid state semiconduction type process. Jager et al. (1987) found a relationship between specific electrical conductivity and viscosity. The rise in the former with temperature was attributed to the decrease in viscosity which facilitates the mobility of the charge carriers. A correlation was observed which was consistent with the charge carriers being concentrated in or on the surface of QI material. Sharp (1987) followed the change in resistivity with temperature. Significant changes were observed in the mesophase region. When the mesophase became the continuous phase, the temperature dependence of the resistivity was sharply reduced, indicating a change in conduction mechanism and the onset of semiconducting type behaviour. Although the studies to date are somewhat limited, the measurement of resistivity would seem worthy of further study and might form the basis of a method of monitoring the changes taking place within the mass of a green carbon structure during the early stages of carbonization.

References

Abson, G. and Burton, C. (1964). Physical tests and range of properties. "Bituminous Materials", Ed. A.J. Hoiberg, Vol. **1**, Interscience (New York, London and Sydney), 213-289.

Barr, J.B. and Lewis, I.C. (1982). Characterisation of pitches by DSC and thermomechanical analysis. Thermochimica Acta **52**, 297-304.

Bartle, K.D., Collins, G., Stadelhofer, J.W. and Zander, M. (1979). Recent advances in the analysis of coal-derived products. J. Chem. Tech. Biotechnol **29**, 531-551.

Benn, M., Rand, B. and Whitehouse, S., (1985). Pitch-mesophase-carbon transformation diagrams for a variety of pitches. Extended Abstracts of 17th Biennial Conf. on Carbon (Amer. Carbon Soc.), Lexington, 159-160.

Blackman, L.C.F., (1970). Modern Aspects of Carbon and Graphite Technology. Academic Press (London and New York).

Collett, G.W. and Rand, B. (1978). Thixotropic changes occurring on reheating a coal-tar pitch containing mesophase. Carbon **16**, 477-479.

Ehrburger, P. and Lahaye, J. (1984). Capillary flow of liquid pitch. Fuel **63**, 1677-1680.

Ferry, J.D. (1969). "Viscoelastic Properties of Polymers", Wiley (New York), 2nd Ed., 292.

Fischer, P., Stadelhofer, J.W. and Zander, M. (1978). Structural investigation of coal-tar pitches and coal extracts by 13C nmr spectrometry. Fuel **57**, 345-352.

Fitzer, E. (1987). The future of carbon-carbon composites. Carbon **25**, 163-190.

Fitzer, E., Mueller, K. and Schaeffer, W. (1971). The chemistry of the pyrolytic conversion of organic compounds to carbon. "Chemistry and Physics of Carbon", Vol. **7**, Ed. P.L. Walker Jr., M. Dekker, New York, 237-383.

Fox, T.G., (1956). Bull. Am. Phys. Soc. **1**, 123.

Greenhalgh, E. and Moyse, M.E. (1970). Contact angle of pitch on carbon surfaces. Proc. 3rd Conf. on Industrial Carbon and Graphite (Soc. Chem. Ind.), London 539-549.

Greinke, R.A. (1986). Kinetics of petroleum pitch polymerisation by GPC. Carbon **24**, 677-686.

Greinke, R.A. and O'Connor, L.H. (1980). Determination of molecular weight distributions of polymerized petroleum petroleum pitches by GPC with quinoline eluent. Anal. Chem. **52**, 1877-1881.

Greinke, R.A. and Singer, L.S. (1988). Constitution of coexisting phases in mesophase pitch during heat treatment. Carbon **26**, 665-671.

Hayes, R.A. (1961). The relation between glass transition, molar cohesion and polymer structure. J. Appl. Polymer Sci. **5**, 318-321.

Hildebrand, J.H. and Scott, R.L., (1950). "The Solubility of Non-electrolytes", 3rd Ed. (Rheinhold, New York).

Hutcheon, J.M. (1970). Manufacturing technology of baked and graphitised carbon bodies. "Modern Aspects of Carbon and Graphite Technology", Academic Press (London and New York), 49-79.

Jager, H., Wagner, M.H. and Wilhelm, H. (1987). Interrelation between η and conductivity of pitches. Fuel **66**, 1554-1555.

Karr, J.C. (1978). "Analytical Methods for Coal and Coal Products", Vol. **2**, (Academic Press, New York).

King, L.F. (1978). Analysis of coal-tar binders for electrodes. "Analytical Methods for Coal and Coal Products", Ed. C. Karr Jr., Vol. **2**, Academic Press (New York), 535-587.

Kremer, H.A. (1982). Recent developments in electrode pitch and coal-tar technology. Chemistry & Industry 18 Sept. 1982, 702-718.

Ladner, W.R. and Snape, C.E. (1978). Application of quantitative 13C nmr spectroscopy to coal derived materials. Fuel **57**, 658-662.

Lahaye, J., Aubert, J.P. and Buscailhon, A. (1976). Interaction between a coke and a tar. Influence of the texture of the coke. Proc. 4th London Int. Carbon and Graphite Conf. (Soc. Chem. Ind.), 118-130.

Mack, C. (1966). Physical chemistry. "Bituminous Materials", Vol. **1**, Ed. A.J. Hoiberg (Interscience, New York), 25-121.

McNeil, D. (1966). The physical properties and chemical structure of coal-tar pitch. "Bituminous Materials". Ed. A.J. Hoiber, Vol. **3**, R.E. Krieger, Huntington, New York, 139-216.

Marsh, H., Latham, C.S. and Gray, E.M. (1985). Structure and behaviour of QI material in pitch. Carbon **23**, 555-570.

Monthioux, M., Oberlin, M., Bourrat, X. and Boulet, R. (1982). Heavy petroleum products, texture and ability to graphitise. Carbon **20**, 167-176.

Nazem, F.F. (1982). Flow of molten mesophase pitch. Carbon **20**, 345-353.

Newman, J. (1976). "Petroleum Derived Carbons", M.L. Deviney and T.M. O'Grady, (Eds.) ACS symp. Ser. 21 (Amer. Chem. Soc.) 52-62.

Rand, B. (1985). Carbon fibres from mesophase pitch. "Strong Fibres", Eds. W. Watt and B.V. Perov, Vol. **1** of "Handbook of Composites", A. Kelly and Yu.N. Rabotnov, (Eds.) North-Holland (Amsterdam), 495-576.

Rand, B. (1987). Pitch precursors for advanced carbon materials; rheological aspects. Fuel **66**, 1491-1503.

Rand, B. and Shepherd, P.M., (1980). Glass transformations for some pitch materials. Fuel **59**, 814-816.

Riggs, D.M. and Diefendorf, R.J. (1979). The solubility of aromatic compounds. Extended abstracts of 14th Biennial Conf. on Carbon, Penn. State Univ.(Amer. Carbon Soc.), 147-148.

Riggs, D.M. and Diefendorf, R.J. (1980a). A phase diagram for pitches. Carbon '80 Preprints of 3rd Int. Carbon Conf., Baden-Baden (Deut. Keram. Ges.) 326-329.

Riggs, D.M. and Diefendorf, R.J. (1980b). Factors controlling the thermal stability and liquid crystal forming tendencies of carbonaceous materials. Carbon '80, Preprints of 3rd Int. Carbon Conf., Baden-Baden (Deut. Keram. Ges.), 330-333.

Riggs, D.M. and Diefendorf, R.J. (1983). Solvent extracted pitch precursors for carbon fibre. Extended Abstracts of 16th Biennial Conf. on Carbon (Amer. Carbon Soc.), San Diego, 24-25.

Riley, W.C. (1965), Graphite. "Ceramics for Advanced Technologies", Eds. .E. Hove and W.C. Riley, John Wiley (New York, London and Sydney) 14-76.

Sharp, J.A. (1987). Electrical resistivity of pitch during heat-treatment. Fuel **66**, 1487-1490.

Smith, F.A., Eckle, T.F., Osterholm, R.J. and Stichel, R.M. (1966). Manufacture of coal-tar and pitches. "Bituminous Materials", Ed. .J. Hoiber, Vol **3**, R.E. Krieger, Huntington, New York, 57-116.

Speight, J.G. and Penzes, S. (1978). Electrical conductivity of aromatic and heteroaromatic compounds in solution. Chemistry & Industry (Lond.), 729-731.

Stevens, W.C. and Diefendorf, R.J. (1986). Thermosetting of mesophase pitches. Carbon '86, Proc. 4th Carbon Conf., Baden-Baden, (Deut. Keram. Ges.), 38-41.

Tillmans, H., Peitzka, G. and Pauls, H. (1978). Influence of the QI material in pitch on carbonization behaviour and structure of pitch coke. Fuel **57**, 171-173.

Traxler, R.N. (1964). Rheology and rheological modifiers other than elastomers. "Bituminous Materials", Ed. A.J. Hoiberg, Vol. **1**, Interscience (New York, London, Sydney), 143-213.

Wallouch, R.W., Murty, H.M. and Heintz, E.A. (1977). Ind. Eng. Chem., Prod. Res. Dev. **16**, 325-329.

Whitehouse, S. and Rand, B. (1982). The pitch-mesophase-coke transformation. Carbon '82, Extended Abstracts of 6th Int. Conf. on Industrial Carbon and Graphite (Soc. Chem. Ind.), London, 183-185.

Whitehouse, S. and Rand, B. (1987). Rheology of mesophase pitch from A240. Extended Abstracts of 18th Biennial Conf. on Carbon, Worcester, (Amer. Carbon Soc.), 175-176.

Chapter 4

Kinetics and Catalysis of Carbon Gasification

H. Marsh and K. Kuo

Northern Carbon Research Laboratories, Dept. of Chemistry,
University of Newcastle upon Tyne, Newcastle upon Tyne,
NE1 7RU, U.K.

Summary.

Carbon gasification reactions form the basis of several important industrial processes. This Chapter considers the fundamental science of gasification of carbons including cokes and chars. The distinction is made between chemical and diffusional control of reaction rate and the influences of porosity in the carbon. Rates of gasification, referred to as reactivity parameters, are a function of reacting gas, temperature, pressure, impurity content and structure of the carbon. The use of <u>active</u> and <u>reactive</u> surface areas to normalise gasification rates as a basis for comparison is explained. Mechanisms of reaction of carbon with atomic and molecular oxygen, carbon dioxide, steam, oxides of nitrogen and hydrogen are reviewed and compared. Catalysis of oxidation reactions is assessed in terms of mechanisms involving oxygen-transfer stages and topographical changes causing pitting and channelling. Inhibition of gasification is introduced. The difficulty in using the concept of reactivity in a rigorous comparative way is discussed.

KINETICS AND CATALYSIS OF CARBON GASIFICATION

H. Marsh and K. Kuo

Northern Carbon Research Laboratories, Dept. of Chemistry, University of Newcastle upon Tyne, Newcastle upon Tyne, NE1 7RU, U.K.

1 INTRODUCTION

Carbon gasification reactions form the basis of several important industrial processes. A major part of the world's energy requirements is met by the combustion of coals in electrical power stations, briquettes are burnt in the domestic market, cokes are gasified as anodes for aluminium production, metallurgical cokes are gasified in the blast furnace either in the stack or at the tuyeres and special carbon composites are constructed to withstand oxidation e.g. refractory materials and carbon-fibre composites in rocket exhausts and space stations.

Reactions with oxygen, steam, carbon dioxide and hydrogen during gasification, combustion, liquefaction, hydrogasification and pyrolysis (Table 1) result in carbons being selectively and continuously removed.

$$C_f + O_2 \rightarrow CO_2$$
$$C_f + H_2O \rightarrow CO + H_2$$
$$C_f + CO_2 \rightarrow 2CO$$
$$C_f + 2H_2 \rightarrow CH_4$$

C_f is a carbon atom(s) at a surface, undergoing gasification.

A description of any study of the kinetics of combustion/gasification of carbons by any oxidizing gases must include a specification of the carbon(s) being gasified.

2 THE NATURE OF CARBON SURFACES

The term 'carbon' covers a very wide range of carbonaceous substances, from chars to single-crystal graphite, see Chapter 1. In carbonaceous materials, carbon atoms are usually in hexagonal rings (as in benzene) with these rings joined together, imperfectly, to form polyaromatic lamellar molecules. These molecules, or lamellae, are not small, perfect sheets of graphite, they also contain heteroatoms, e.g. hydrogen, oxygen, nitrogen, sulphur and phosphorus. In addition to chemical defects, crystallographic defects present affect carbon structure. There may be single and multiple atom vacancies in the lamellar molecules, dislocations, grain boundaries, stacking faults and defects, e.g. point defects, basal and screw dislocations (Fischbach 1971). Surface contamination (inorganic mineral matter) is often catalytically active during gasification. These imperfect lamellar molecules are the "building-bricks" of structure within carbonaceous materials. (Chapter 1).

Such materials can be classified into graphitizable carbons and non-graphitizable carbons (Kochling et al. 1982 and Marsh 1979). The essential difference between them is the dimension of structural order associated with these lamellar molecules. The definitions of non-graphitizable and graphitizable carbon are given in Chapter 1.

The structure of any carbon surface depends upon the structure of the organic compound(s) in the precursor and the thermal treatment given to the precursor. Due to the heterogeneous nature of the carbon lamellae, an oxidizing gas molecule approaching a surface of carbon is presented to carbon atoms in diverse structural environments.

3 REACTIVITY OF CARBONS

Different carbons exhibit different reactivities to oxidizing gases. Generally, the reactivities of graphitizing (Marsh et al. 1981, Smith et al. 1956) and non-graphitizing carbons (Board 1966) to oxygen and carbon dioxide decrease with increasing heat treatment temperature of carbon preparation. Thermal annealing due to heat-treatment (van Krevelen 1961, Franklin 1949, Blake et al. 1967) results in the reduction of structural defects thereby reducing carbon reactivity. Jenkins et al. (1973) found, for example, that lower-rank coal chars (non-graphitizing) are more reactive than anisotropic cokes prepared from high rank coals. Carbons containing mineral matter impurities (Hippo et al. 1975, Walker et al. 1968 and Rakszawski 1963) are more reactive.

3.1 Selective gasification

Topographically, carbons do not gasify evenly. Selective gasification occurs at particular sites while other sites remain unaffected. Such active or favoured sites are located mainly at the prismatic edges of the carbon layers (Hennig 1966, Thomas, 1965), and at defects, e.g. vacancies or dislocations (Ergun et al. 1965) in the basal plane. Inorganic impurities, by promoting catalytic activity, create further dislocations. Inorganic impurities are often located at defect centres. It is suggested by Laine et al. (1963a) that heteroatoms control the initial site of reaction.

Variations in reactivity of 'pure' carbons can be accounted for by differences in the number of active sites per unit surface area and the total surface area (TSA) accessible to the reacting gas molecule. The active surface area (ASA) is a measure of the available surface area which takes part in the initial reaction.

Edge carbon atoms are more reactive than basal plane carbon atoms (Stein et al. 1985) and are influenced by the presence of catalytic impurities (Thomas et al. 1967, Pinnick 1956). Geometrically, edge carbon atoms can readily form bonds with chemisorbed oxygen due to the availability of unpaired sp^2 electrons. Basal plane carbon atoms presumably have their π-electrons forming chemical bonds with adjacent carbon atoms.

There are differences also between different structures of edge atoms. Stein and Brown (1985) using structure-resonance theory (Swinborne-Sheldrake et al. 1975) on large benzenoid polyaromatic molecules with different edge structures (Figure 1) found that molecules with "zig-zag" edges (A) are more reactive than molecules with "arm-chair" edges (B). By applying first-order perturbation theory to the Huckel molecular orbitals (Heilbronner et al. 1976), They estimated electron localization energies (δE_π). These are the energies required to isolate a π-electron from the rest of the conjugated system for polyaromatic molecules of different sizes and edge structures. They found that molecules with "zig-zag" edges (A) are more reactive than molecules with "arm-chair" edges (B) (Figure 1). Previous workers (Lewis et al. 1963, Madison et al. 1963) also supported these findings. Catalytic impurities (Walker et al 1959, Tomita et al. 1977) are effective on reacting surfaces and can be mobile.

Figure 2 shows an oxygen molecule approaching a graphite lattice where it could undergo reaction either at the edge of the basal plane (prismatic edge) or on the basal plane. Rates of gasification at the prismatic edges are 10^2 to 10^3 times faster than on the basal plane. The diagram illustrates a fact of significance in the carbon molecular oxygen reaction, namely that for the product CO_2, (O-C-O), to be formed, the oxygen molecule must dissociate initially into atoms.

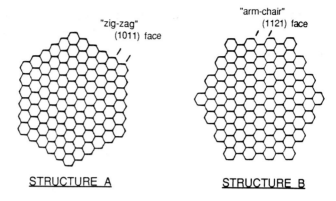

Figure 1. Large benzenoid polyaromatic molecules with :
(A) "zig-zag" edge structure, and
(B) "arm-chair" edge structure.
(Stein and Brown, 1985)

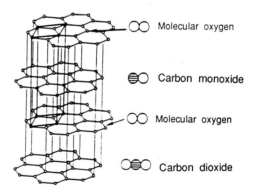

Figure 2. Diagram illustrating the approach of oxygen molecules to the basal plane and prismatic edge of the graphite lattice to form carbon monoxide and carbon dioxide.

The heteroatom oxygen, especially if carbonyl or heterocyclic, is said to promote reactivity via electron exchange (Blackwood et al. 1959). Chemisorbed hydrogen is said to enhance reactivity by providing sites of preferential oxidation with subsequent production of highly reactive 'nascent' carbon sites (Phillips et al. 1970). The availability of π-electrons at heterocyclic sites (e.g. nitrogen and sulphur) could also favour ring structure attack.

4 REACTION KINETICS AND MECHANISMS
4.1 Chemical and diffusion control of rate

As carbon gasification reactions are sensitive to mass transfer effects, it is essential to consider the implication of these before embarking on a discussion of kinetics and mechanisms.

The reactions of carbon with oxidizing gases are controlled by the following processes:-

(1) Mass Transfer (by diffusion) of gaseous reactant(s) from the bulk gas phase to the carbon surface.
(2) Adsorption of reactant(s) on the surface.
(3) Occurrence of chemical rearrangements (reactions) on the surface mobility and formation of adsorbed product(s).
(4) Desorption of product(s).
(5) Mass Transport (by diffusion) of the gaseous product(s) away from the carbon surface.

Step (3) can itself involve several processes. The kinetics of the reaction will be governed by whichever of these processes is the slowest, i.e. "the rate determining step" (RDS).

The RDS is dependent upon process parameters and carbon properties, e.g.
(1) Process parameters:-
    (i)   Temperature
    (ii) Pressure
    (iii) Particle size.

(2) Carbon properties:-
    (i)   Porosity
    (ii) Active site concentration
    (iii) Catalytic impurities.

For the gasification of a non-porous carbon, two distinct situations can be identified. At a 'high' reaction temperature, the RDS is transport of reactant gas molecules to the carbon surface i.e. diffusional control. At lower reaction temperatures, the RDS becomes one of chemical control.

In the case of a porous carbon, a further possibility arises. The reaction may be controlled by chemical processes over the accessible external surfaces but controlled by diffusion in the less accessible internal surface porosity (Figure 3) because of restrictions on the diffusion of gaseous reactants and products into and out of the pores, particularly micropores. This usually provides an intermediate regime occurring in a temperature range between that where pure chemical or diffusional control of the rate applies.

Gasification processes can therefore be divided into three temperature zones as illustrated in Figure 4 with an idealized Arrhenius plot and the corresponding variations in the steady state reactant gas concentrations through the sample.

Zone I Reaction

At low temperatures where the reaction rates are low and are chemically controlled over the whole of the accessible surface, the reactant gas concentration is uniform throughout the bulk of the carbon sample and equals the concentration in the gas phase. The measured activation energy in this zone is the true chemical activation energy of the reaction.

Only if the reaction is occurring in this zone can unequivocal information on the mechanism of the gasification reactions be obtained from the experimental kinetic data. Low temperatures and pressures and small particle size (if the sample is porous) are required to obtain a reaction in Zone I.

Zone II Reaction

In this temperature region, the rate is partially controlled by diffusion in the pores so that the gaseous reactant concentration falls gradually to zero at some point in the pore system.

Zone III Reaction

At high temperatures, the reaction rate is controlled only by diffusion of the gas to the external carbon surface. A boundary layer exists within which the gas concentration is depleted from its bulk value. The activation energy is lower than that of Zone II. In practice, a transition region between each of the zones is often observed with deviation from the Arrhenius plot of the chemical reactions.

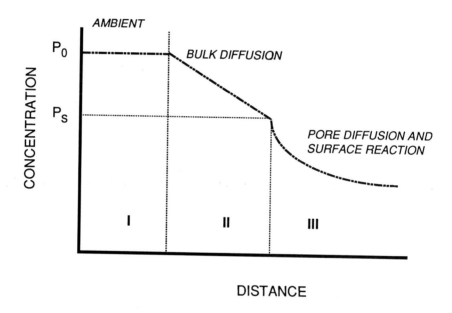

Figure 3. Concentration profile for gaseous reactant near and within a porous char particle.

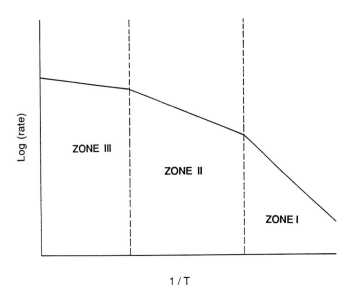

Figure 4 (a). Diagram showing the change in rate of the carbon gasification reaction with temperature (Walker, Rusinko, and Austin, 1959).

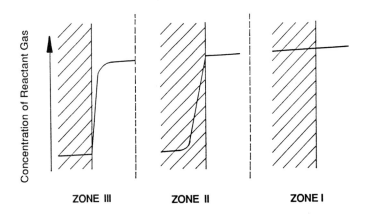

Figure 4 (b). Diagram showing how the concentration of reactant gas in the specimen varies with rate controlling zone (Walker, Rusinko, and Austin, 1959).

4.2 Reaction rates

The overall chemical kinetics of heterogeneous carbon reactions with oxidizing gases is calculated by the following equation:-

$$R = - \frac{1}{W} \frac{dW}{dt} \qquad (1)$$

where R is the overall reactivity at temperature T K ($g\ g^{-1}\ s^{-1}$), W = initial mass of the carbon (g), and dW/dt = maximum rectilinear rate of weight loss. The intrinsic reactivity expressed per unit surface area is usually given by:-

$$R_i = k_i\ P^m\ A_g^{-1}\ (g\ m^{-2}\ s^{-1}\ Pa) \qquad (2)$$

where R_i = Intrinsic reactivity; k_i = Intrinsic rate coefficient, P = partial pressure of reactant gas and m = true reaction order, A_g = surface area ($m^2\ g^{-1}$).

The rate coefficient, k, is related to temperature via the Arrhenius expression:-

$$k = A\ e^{-E/RT}$$

Where A is the pre-exponential factor and E the activation energy.

Several mechanisms have been proposed to explain carbon-gas reactions and the relationship between overall (Equation 1) and intrinsic reactivity (Equation 2).

4.3 Chemisorption and desorption

According to active site theory, reactions occur only at active or favoured sites. As discussed above, these active sites are (i) carbon edges or defects (ii) heteroatoms (iii) inorganic impurities. The gas molecule is adsorbed, reacts to form an intermediate and desorbs to produce gaseous products plus a free site (Figure 5). Both adsorption and desorption occur via a single site or dual site mechanism.

Single Site Mechanism (SSM) requires one free carbon site which can lead to the simultaneous production of gaseous species, *e.g.*

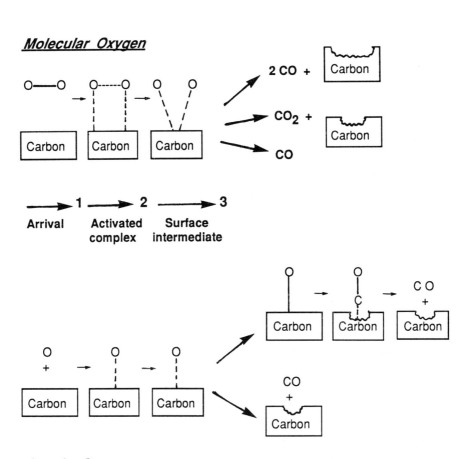

Figure 5. Schematic representation of events in molecular and atomic oxygen adsorption and film decomposition.

$$CO_2 + C_f \rightarrow C(O) + CO$$

$$O + C_f \rightarrow C(O)$$

where $C(O)$ denotes a carbon site with chemisorbed atomic oxygen and C_f denotes a free carbon site.

Dual site mechanism (DSM) requires two free active sites e.g. $O_2 + 2C_f \rightarrow 2C'(O)$ which may migrate to a new site to form a more stable, immobile surface intermediate, $C(O)$ or vice versa.

$$C'(O) \Leftrightarrow C(O)$$

The surface intermediates may undergo either a single or dual site desorption e.g.

$$C(O) \rightarrow CO + C_f$$

$$C'(O) + C'(O) \rightarrow CO_2 + C_f$$

The active site theory assumes:-
(1) localized adsorption via collisions with active sites,
(2) one molecule or atom is adsorbed per active site,
(3) a constant surface mechanism (chemisorption/migration/desorption),
(4) the surface coverage is less than a complete monolayer.

Langmuir-Hinshelwood kinetics (Hayward et al. 1964, Frank-Kamenetskii 1969), using active site theory, describe the carbon-gas reaction in terms of chemical adsorption mechanism. In this mechanism, the above assumptions are used together with the following three assumptions:-
1. The surface is homogeneous with a uniform distribution of active sites over the entire surface.
2. There are no interactions between adsorbed species. The adsorption rate per site is not affected by the adsorbed species.
3. There is either no or very rapid surface migration so that only adsorption and desorption can be rate controlling.

The Langmuir-Hinshelwood kinetics assume a homogeneous, non-interacting surface thus implying that both the activation energy for adsorption E_A and desorption E_d remain constant in time, as well as from site to site.

Assuming steady state, isothermal conditions,

$R_a = R_d$

where R_a and R_d are intrinsic rates of adsorption and desorption respectively. If the intrinsic reactivity, R_i, is controlled by either the rate of adsorption or desorption, then using Langmuir-Hinshelwood kinetics,

$$R_i = R_a = R_d \qquad (3)$$

where $R_a = k_a P \left(1 - \dfrac{aP}{1+aP}\right) \qquad (4)$

and $R_d = k_d \left(\dfrac{aP}{1+aP}\right) \qquad (5)$

k_a or k_d = rate constant for adsorption or desorption
a = a constant dependent on temperature
P = partial pressure of the reactant gas.

Comparing Equations 2 and 3, the true order of the reaction will fall in the range $0 \leqslant m \leqslant 1$ dependent on the constant 'a' and range of partial pressure 'P' interest:

(1) $aP \ll 1 \rightarrow R_i = k_a P$ [m=1]
(2) $aP \gg 1 \rightarrow R_i = k_d$ [m=0]

Modifications are made to the Langmuir-Hinshelwood mechanism by taking into account multicomponent systems where sites on the reacting surface may be taken up by inert, reactant or product gases thus inhibiting the reaction (Frank-Kamenetskii 1969).

For a non-homogeneous surface, the kinetics can be developed assuming that most active sites react first, especially at higher temperatures where surface mobility is enhanced. For an interacting surface, filling of nearby sites creates repulsion forces thus inhibiting adsorption (and promoting desorption) of following molecules. Consequently E_a increases (and E_d decreases) as the surface coverage (O) increases.

$$E_a = E_{ao} + w_a \qquad (6)$$

$$E_d = E_{do} - w_d \qquad (7)$$

where E_{ao} and E_{do} are activation energies at $\theta = 0$; w_d and w_a are surface constants.

The chemisorption rate on carbon often follows Equation 6 via the 'so-called' Elovich equation (Hayward et al. 1964).

$$R_a = R_{ao} \, e^{-w_a \theta / RT} \qquad (8)$$

where R_{ao} is the adsorption rate at $\theta = 0$.

4.4 Importance of active surface area (ASA) to reactivity

The intrinsic reactivity, R_i (Equation 2) for a heterogeneous carbon surface can be expressed in another way to account for the active sites:-

$$R_i = k_i \, [C_t] \, W \, P^m \qquad (9)$$

where k_i = intrinsic rate coefficient, $[C_t]$ = active site concentration, W = mass of carbon atom, P = partial pressure of the reactant gas, and m = true reaction order.

Often, the intrinsic reactivity, R_i, is reported relative to the external surface reactivity, R_s, where $P = P_s$ (Figure 2):-

$$R_i = R_s \text{ and} \qquad (10.\text{i})$$
$$R_s = k_i [C_t] W \, (P_s)^m \text{ where} \qquad (10.\text{ii})$$
P_s = partial pressure of gas at the surface.

The overall reactivity R is usually related to the intrinsic reactivity, R_i (= R_s) by

$$R = \eta A_g R_s \text{ where} \qquad (11)$$
η = degree of gaseous penetration and
A_g = total internal surface area ($m^2 \, g^{-1}$).

Combining Equations 10 and 11 and assuming an isothermal sample,

$$R = k_i \eta A_g [C_t] W (P_s)^m \text{ where} \tag{12}$$
$[C_t]$ = active site concentration.

Equation 12 thus implies that $R \alpha [C_t] A_g$ i.e. carbon reactivity is proportional, not to total surface area (TSA) but to active surface area (ASA).

Laine et al. (1963a) verified this conclusion by direct measurement of ASA on graphitized carbon black. Oxygen was chemisorbed on a 'clean' carbon surface at 300°C for 24 h, the resultant surface oxygen complex being desorbed as CO/CO_2 at a higher temperature. The amount of surface oxygen complex formed is a measure of ASA, assuming one oxygen atom per carbon atom and a specific surface area of 0.083 nm^2 per site. Radovic et al. (1983) and Garcia et al. (1986) extended the concept of active sites to lignitic chars and observed that reactivity values could be better normalized when the ASA of the char is considered. Jenkins and Piotrowski (1978) carried out a study of the degree of gasification (burn-off) on changes in active site concentration. The concept of "reactive" surface area (RSA) was introduced, calculated and utilized to normalize reaction rates in terms of "turn-over number" (TON). The reactive surface area is measured in terms of extents of oxygen desorbed from the surface at reaction temperature when the reacting gas is replaced, e.g. by nitrogen or vacuum. More stable chemisorbed oxygen does not relate to reactivity. Values of TON were relatively constant for several chars over a wide range of reaction temperatures. Khan (1987) studied possible influences of hydrogen present on char surfaces on ASA determinations. Normalization of these reactivity results can be improved further when the measured active sites are corrected for the hydrogen present on the char surface. Smith (1978) collected published data of intrinsic reactivities of a large variety of carbons and compared them. Such values of the intrinsic reaction rates per unit TSA, at a given temperature and otherwise constant conditions, differ by up to four orders of magnitude (Figure 6). The variations can be accounted for, in part, by differences in heat treatment temperature and impurity levels of the carbon.

4.5 Concept of reactivity

Whereas such properties of a carbon as its temperature or mass can be evaluated accurately, the 'reactivity' of a carbon cannot be given such precision.

The intrinsic reactivity R_i of carbon is a function of the gas used. It does not follow that for several carbons, reactivity will be in the same order for all gasifying gases. Therefore, intrinsic reactivity is always a comparative value in terms of application

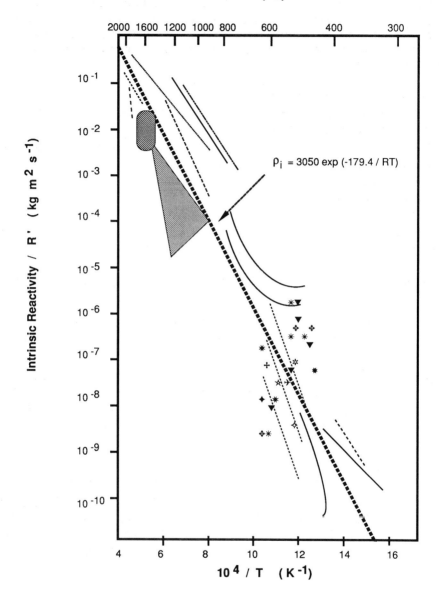

Figure 6. Variation of intrinsic reactivity of various carbons with temperature and oxygen pressure of 101 kPa (based on Smith, 1978).

and use. Units of R_i are g m^{-2} s^{-1} Pa, thus, a surface area evaluation is necessary. The measurement of the surface area available to a reacting gas molecule is not, however, a straightforward task. Methods of physical adsorption of gases to obtain isotherms and their interpretation often indicate pore volume rather than surface area.

In summary, carbon reactivity is influenced by:
(a) parent material: e.g. cellulose to polycyclic hydrocarbons, giving a more or less defective constituent lamellar molecule structure,
(b) the heat treatment temperature (HTT) of the carbon, increasing temperature decreasing defect density and decreasing reactivity,
(c) accessible surface area,
(d) catalysis by inorganic impurities.

The concept of Active Surface Area, which attempts to rationalise differences in carbon reactivity, can be further described using Figures 7, 8 and 9.

All Figures are plots of a reactivity function against total surface area (TSA) of carbons. TSA can range from <0.1 m^{2}g^{-1} to >1000 m^{2}g^{-1}. Generally, for carbons of HTT of 900°C, low surface area carbons are anisotropic (graphitizable) and high surface area carbons are isotropic (non-graphitizable).

Figure 7 uses the overall reactivity R(s^{-1}) plotted against the measured accessible total surface area (TSA) of the carbon. If all surfaces are comparable in terms of reactivity, then there should be a linear relationship between R and TSA (solid line of Figure 7). However, surfaces could be more or less reactive (broken lines) or could be randomly distributed.

Figure 8 uses the intrinsic reactivity R_i' (g m^{-2} s^{-1}) based on TSA plotted against the measured accessible total surface (TSA) of the carbon. If all surfaces are comparable in terms of reactivity then they should all have the same reactivity (horizontal line of Figure 8). Differences between surfaces will be indicated by deviations from this line. For example, significant catalysis will give data points above the horizontal line.

Figure 9 uses the intrinsic reactivity R_i" (g m^{-2} s^{-1}) based on ASA plotted against the measured accessible total surface area (TSA) of the carbon. If all surfaces have equal densities of active sites (and hence reactivity) then they all should have the same reactivity value (horizontal line of Figure 9). Again, differences between the active sites of carbons will be indicated by deviations from this line. Significant catalysis will give data points above the horizontal line.

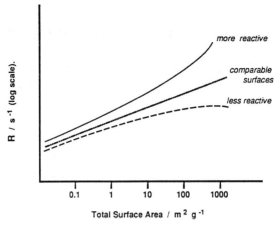

Figure 7. Carbon gasification. **R** vs **TSA**.

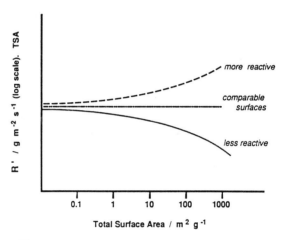

Figure 8. Carbon gasification. **R'** vs **TSA**.

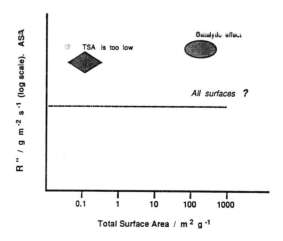

Figure 9. Carbon gasification. **R"** vs **TSA**.

5 THE CARBON-MOLECULAR OXYGEN REACTION

Despite extensive investigations, the mechanism of the carbon-oxygen reaction is not understood in detail. The reasons for this are because the reaction is extremely sensitive to the limitations imposed by energy and mass transfer processes. Consequently, kinetic studies are often studied at low temperatures and pressures, i.e. conditions which are far removed from many industrial operations. The activation energy of this exothermic reaction in the Zone 1 temperature region is 230-270 kJ mol^{-1} (Walker et al. 1959). Two gaseous products viz. carbon monoxide and carbon dioxide are formed when molecular oxygen reacts with carbons:-

$$2 C_f + O_2 = 2CO \tag{13}$$

$$C_f + O_2 = CO_2 \tag{14}$$

where C_f = free carbon site.

There is evidence that both CO and CO_2 are primary products (Walker et al. 1959), with the CO/CO_2 ratio increasing substantially at higher temperatures and lower pressures (von Fredersdorff et al. 1963). A possible explanation is that CO is formed at carbon edges while CO_2 is formed at inorganic sites. Lower temperatures favour CO_2 due to catalytic activity; higher temperatures promote utilization of carbon edges. The CO-CO_2 ratio can be expressed (Phillips et al. 1970; Arthur 1956) as:-

$$CO/CO_2 = A\, e^{-E/RT}$$

where $A = 10^{2.5}$, E = 25-38 kJ mol^{-1} at low pressures and $A = 10^{3.5}$, E = 50-79 kJ mol^{-1} at high pressures.

The surface oxygen complexes are involved in carbon monoxide and carbon dioxide formation. These are either stable chemisorbed oxygen species or reaction intermediates. The complex may either be localized or mobile over the carbon surface.

The following are suggested reaction stages to describe the mechanism of formation of carbon monoxide and carbon dioxide:-

$$C_f + O_2 \rightarrow C(O_2) \text{ or } C(O_2)_m \tag{15}$$

$$C(O_2)_m \rightarrow C(O) + C(O)_m \text{ or/and} \tag{16}$$

$$\rightarrow C(O)_m + C(O)_m \text{ or/and} \tag{17}$$

$$\rightarrow C(O) + C(O). \tag{18}$$

$$C(O) \rightarrow CO \tag{19}$$

$$C(O)_m \rightarrow CO \tag{20}$$

$$C(O)_m + C(O)_m \rightarrow C_f + CO_2 \tag{21}$$

$$C(O)_m + C(O) \rightarrow C_f + CO_2 \tag{22}$$

$$CO + C(O) \rightarrow C_f + CO_2 \tag{23}$$

$$CO + C(O)_m \rightarrow C_f + CO_2 \tag{24}$$

$$O_2 + 2\,C(O) \rightarrow 2\,CO_2. \tag{25}$$

where C_f = free carbon site

$C(O_2)$ = chemisorbed, localized molecular oxygen

$C(O_2)_m$ = chemisorbed, mobile molecular oxygen

$C(O)$ = chemisorbed, localized atom of oxygen

$C(O)_m$ = chemisorbed, mobile atom of oxygen.

Knowledge of the role of the surface oxygen complex in the mechanism of gasification comes largely from measurements by Walker and co-workers (Laine et al. 1963a; Vastola et al. 1964; Lang et al. 1962).

The reaction proceeds at low pressures at a measurable rate at temperatures above about 600 K, producing carbon monoxide and dioxide from both mobile and localized reactive surface oxygen complexes (Laine et al. 1963a, 1963b; Rossberg 1956; Wicke 1955; Day et al. 1958). (Note, at higher oxygen pressures of e.g. 20kPa, the carbon would ignite and combust. No kinetics are possible in this regime). The residence time of these complexes may be relatively short. Stable surface oxide which could be either mobile or localized is also produced and is the only product at temperatures below 573 K. Vastola et al. (1964), using both oxygen-16 and oxygen-18 isotopes, found a slow decomposition of the relatively stable oxygen complex. Ahmed and Back (1985) using high purity thin carbon films to measure the formation of surface oxide complex and CO/CO_2 products as a function of time, together with comparison of the rates of reaction in the presence and absence of complex, showed that the complex is an intermediate in the gasification process.

A basic rate equation proposed by Walker et al. (1959) is:-

$$\frac{-d[C_g]}{dt} = \frac{k_c(P_{O_2})^n ASA (1-\theta)}{[1-\pi(\theta)]} \tag{26}$$

where $d[C_g]/dt$ = rate of total carbon gasified
$\equiv$ rate of oxygen depletion
k_c = the rate constant
P_{O_2} = partial pressure of oxygen
n = order of the reaction with respect to oxygen
ASA = Active Surface Area
θ = fraction of the ASA covered with complex
$\pi(\theta)$ = participation factor reflecting the contribution of the complex in producing CO and CO_2.

The gaseous CO and CO_2 are considered to be produced by two paths:-

(i) via a fleeting oxygen intermediate,

(ii) via the breakdown of a more stable surface oxide complex.

The denominator $[1-\pi(\theta)]$ in Equation 26 takes into account the contribution to the rate, $-d[C_g]/dt$, of the slow decomposition of the oxide.

Several studies report examples of a reaction order $n = 1$ (Walker et al. 1959); recent studies (Lewis 1970; Rodriguez-Reinoso 1974) report values for the order of reaction in the range 0.5 to 0.6 at pressures up to 0.1 MPa and temperatures up to 1023 K. As measured order relates to extent of coverage of active surface, and thus is a sensitive function of surface structure and carbon origin, close agreement between unrelated studies is not anticipated.

The reactive 'transient' surface intermediate and the stable surface oxide complex form only on the active sites, the total area of which is termed the ASA, this being only a fraction of the total surface area (Vastola et al. 1964; Laine et al. 1963b).

6   THE CARBON-CARBON DIOXIDE REACTION

The overall reaction of carbon with carbon dioxide (the Boudouard reaction) is:-

$$C_{(s)} = CO_2 \rightarrow 2CO_{(g)} \tag{27}$$

The mechanism and kinetics of this reaction have been extensively studied (Gadsby et al. 1948; Reif 1952; Ergun 1956; Ergun et al. 1965). There is general agreement that the gasification of carbon by carbon dioxide at pressures up to 0.1 MPa can be described by a rate equation of Langmuir-Hinshelwood form:-

$$\text{Rate} = \frac{k_1 P_{CO_2}}{1 + k_2 P_{CO} + k_3 P_{CO_2}} \tag{28}$$

where P_{CO_2}, P_{CO} are the partial pressures of carbon dioxide and carbon monoxide respectively.

k_1, k_2, k_3 are constants which are a function of the rate constant of the individual steps of the reaction (see below).

It is apparent from Equation 28 that no simple order or activation energy can exist. However, under certain conditions, the reaction can be simplified. At low temperatures and high carbon dioxide pressures, the rate becomes equal to k_1/k_3 and should be zero order. On the other hand, at low temperatures and low carbon dioxide pressures, the rate is approximately equal to $k_1 P_{CO_2}$, i.e. first order. Experimentally, reaction orders vary from zero to one dependent upon the temperature, pressure and the nature of the carbon reacted.

Two mechanisms have been proposed both of which give a rate equation of the form of Equation 28:-

Mechanism A

$$C_f + CO_{2(g)} \xrightarrow{i_1} C(O) + CO_{(g)} \quad (29)$$

$$C(O) \xrightarrow{j_3} CO_{(g)} + nC_f \quad (30)$$

$$C_f + CO_{(g)} \underset{j_2}{\overset{i_2}{\rightleftarrows}} C(CO) \quad (31)$$

where i_1, j_3, i_2 and j_2 are the rate constants for those reactions. At steady state, where the rates of formation and removal of the surface complexes are equal, $k_1 = i_1$, $k_2 = i_2/j_2$ and $k_3 = i_1/j_3$.

Mechanism B

$$C_f + CO_{2(g)} \underset{j_1}{\overset{i_1}{\rightleftharpoons}} C(O) + CO_{(g)} \quad (32)$$

$$C(O) \xrightarrow{j_3} CO_{(g)} + C_f \quad (33)$$

where C_f represents a vacant active site

$C(O)$ represents an active site occupied by an adsorbed oxygen atom

$C(CO)$ represents an active site occupied by an adsorbed carbon monoxide molecule, and

$k_1 = i_1$, $k_2 = j_1/j_3$ and $k_3 = i_1/j_3$.

The essential difference between the two mechanisms lies in the explanation for the retardation of the reaction by carbon monoxide. Both mechanisms state that the retardation results from a reduction in the fraction of the active surface covered by oxygen atoms.

Mechanism A proposes that carbon monoxide, by chemisorbing on active sites and forming C(CO) complex, reduces the active site area and consequently the gasification rate.

In Mechanism B, carbon monoxide inhibits the gasification rate by the back reaction with the C(O) complex (Equation 32) thereby reducing the C(O) complex on the surface and hindering further formation of CO by Equation 33.

Mechanism B is now generally favoured (Thomas 1970) largely due to Reif (1952) who showed that the chemisorption of carbon monoxide is negligible and cannot account for the retardation. Equilibrium is rapidly established for Equation 32 and Equation 33 is rate controlling with the rate of reaction proportional to the concentration of occupied active sites.

Inconsistency was noted by Strange and Walker (1976) with the single-site Langmuir-Hinshelwood model proposed by Ergun (1956). Koenig et al. (1985) proposed a model involving a two-site adsorption and dissociation of CO_2:-

$$CO_2 + 2C_f \Leftrightarrow C^* \tag{34}$$

where C^* is the two-site surface complex. A possible structure for C^* is a lactone-type surface species:-

Barton and Harrison (1975) provided evidence for lactone-type structures on graphite and Spheron-6 (a carbon black). The lactone-type species can eventually break down to re-form reactants or dissociate as follows:-

$$C^* \Leftrightarrow C(O) + C(CO) \tag{35}$$

7 THE CARBON-STEAM REACTION

The gasification of carbon by steam to form hydrogen and carbon monoxide is important in many industrial processes such as coal gasification, activation of carbon, regeneration of coked catalysts, water-gas manufacture and in nuclear technology.

Ergun and Mentser (1965) described the carbon-steam reaction by a similar rate equation to that of the carbon-carbon dioxide Equation 28:-

$$\text{Rate} = \frac{k_1 P_{H_2O}}{1 + k_2 P_{H_2} + k_3 P_{H_2O}} \tag{36}$$

where P_{H_2O}, P_{H_2} are the partial pressures of water and hydrogen respectively; k_1, k_2 and k_3 are functions of one or more rate constant. The overall reaction is:-

$$C + H_2O \Leftrightarrow CO + H_2 \tag{37}$$

However, the products can react further, the final gas composition of the products depending on experimental conditions. The following secondary reactions which are important are:-

$$CO + H_2O \Leftrightarrow CO_2 + H_2 \text{ (Water-gas shift reaction)} \tag{38}$$

$$2CO \Leftrightarrow C + CO_2 \text{ (Boudouard reaction)} \tag{39}$$

Two mechanisms have been proposed to account for the rate equations above (Equation 36).

Mechanism A

$$C_f + H_2O_{(g)} \rightarrow C(O) + H_2 \tag{40}$$

$$C(O) \rightarrow CO_{(g)} \tag{41}$$

$$H_{2(g)} + C_f \Leftrightarrow C(H_2) \tag{42}$$

Mechanism B

$$C_f + H_2O_{(g)} \Leftrightarrow C(O) + H_2 \tag{43}$$

$$C(O) \rightarrow CO_{(g)} \tag{44}$$

where C_f represents a vacant active site,
$C(O)$ represents an active site occupied by an adsorbed oxygen atom,
$C(H_2)$ represents an active site occupied by an adsorbed hydrogen molecule.

Mechanism A predicts inhibition by hydrogen adsorption on active sites (Equation 42) whilst Mechanism B proposed by Walker et al. (1959) predicts inhibition by removal of the C(O) complex by hydrogen (Equation 43).

No clear distinction has been made as to which mechanism is correct. Walker et al. (1959) consider, as for the carbon-carbon dioxide reaction, that the rate controlling process is the desorption of the product. The activation energy in the Zone 1 temperature region is 270-310 kJ mol^{-1}.

8   THE CARBON-OXIDES OF NITROGEN REACTION

Oxides of nitrogen which are formed during coal combustion can subsequently react with carbon to produce nitrogen gas (Smith et al. 1957; Bedjai et al. 1958; Watts 1958).

$$C + NO \rightarrow CO + \tfrac{1}{2}N_2 \quad (45)$$
$$C + 2NO \rightarrow CO_2 + N_2 \quad (46)$$
$$C + N_2O \rightarrow CO + N_2 \quad (47)$$
$$N_2O + CO \rightarrow N_2 + CO_2 \quad (48)$$

The reaction of carbon with nitrous oxide is intermediate in rate between that of oxygen and carbon dioxide (Madley et al. 1953). The activation energy of wood charcoal-nitrous oxide reaction is 134 kJ mol^{-1}. The above reaction (Equation 48) at 'low' temperatures (<327 K) occurred only as a catalytic reaction on the charcoal surface.

9   THE CARBON-HYDROGEN REACTION

There have been relatively few kinetic studies of the reaction of carbon with hydrogen (Zielke et al. 1955; Feistel et al. 1977). High hydrogen pressures were employed but even under these conditions, the rate is relatively slow.

Zielke and Gorin (1955) (T = 1083-1201 K, P ≈ 3MPa) and Feistel et al. (1977) (T = 873-1373 K, P = 1 to 7 MPa) propose the following rate equation for the conversion of carbon and hydrogen to methane:

$$\text{Rate} = \frac{k_1 (P_{H_2})^2}{1 + k_2 (P_{H_2})} \quad (49)$$

where P_{H_2} is the partial pressure of hydrogen and k_1 and k_2 are functions of one or more rate constants.

The rate Equation 49 can be explained by the following mechanism:-

$$H_2 + C_f \Leftrightarrow C(H_2) \quad (50)$$

$$C(H_2) + C_f \Leftrightarrow 2C(H) \quad (51)$$

$$2C(H) + H_2 \Leftrightarrow 2C(H_2) \quad (52)$$

$$2C(H_2) + 2H_2 \Leftrightarrow 2CH_4 + 2C_f \quad (53)$$

where C_f represents a surface methylene (-CH=CH-)
$C(H_2)$ represents an active intermediate

The existence of C(H) was postulated because the reaction between carbon and atomic hydrogen also produces methane (Blackwood et al. 1959) The reaction overall must be complicated and the detail of the mechanism is not really given by Equations 50-53).

10 COMPARISON OF CARBON GASIFICATION REACTIONS

A comparison of carbon reactions with different oxidizing gases is in Table 1.

It is rarely possible to measure experimentally the rates of these different gasification reactions under comparable conditions of temperature and pressure because of the large differences between their relative rates (Table 1). These rates (Table 1) are estimates for particular sets of conditions. Any such estimate depends on the temperature at which the rates are compared because the activation energy in Zone 1 is not the same for the different reactions.

Table 1. <u>Carbon gasification reactions</u> (Walker, et al. 1959)

Reaction	ΔH (kJ mol^{-1})	Relative Rates at 1073 K and 0.1 atm.	Activation Energy in Zone I (kJ mol^{-1})
$C + CO_2 \rightarrow 2CO$	170.7	1	335-375
$C + H_2O \rightarrow CO + H_2$	130.3	3	270-310
$C + O_2 \rightarrow CO_2$	-393.4	1×10^5	230-270
$C + 2H_2 \rightarrow CH_4$	-74.8	3×10^{-3}	150

A wide range of activation energies of carbon gasification reactions has been reported, from 40-400 kJ mol^{-1}. Many reported low values for the activation energy are erroneous due to diffusional and catalytic influences. The surface oxygen complexes formed at different temperatures vary in extents of surface coverage and energetics of bonding. As a result, activation energies differ with different gasifying gases.

11 CATALYSIS OF OXIDATION REACTIONS

Inorganic impurities present in carbonaceous materials usually increase catalytically or may sometimes decrease gasification rates. In some uses of carbonaceous solids, catalysed gasification reaction is undesirable (Morgan et al. 1982); in others, it is a desired reaction e.g. catalytic coal gasification enables a one-step process for the conversion of coal to methane (Nahas 1983; McKee 1983).

Many metals, the most active being transition metals, alkali metals and alkaline earth metals, exhibit catalytic activity (Walker et al. 1968).

The extent by which a particular catalyst will accelerate gasification rate is a complex function of many variables including:-

  (i)   the metal concerned,
  (ii) the gasification reaction being studied and thermal conditions employed,
  (iii) the size of the catalyst particles and their state of dispersion throughout the carbon,
  (iv) the chemical state of the catalyst, and
  (v)   the relative amounts of catalyst.

In the majority of earlier investigations, not all of these important parameters were recognized. This is a major reason for the lack of agreement between workers of relative activities of catalysts and activation energies for the catalysed reactions.

11.1 Effects of catalysts on reaction kinetics

Catalysts usually provide a new reaction pathway, the slow stage of which has a lower energy of activation than the uncatalyzed reaction (Walker et al. 1968; McKee 1981). The decrease in activation energy of the rate determining step of the reaction results in an increase in reaction rate.

The change in activation energy, E, of a reaction due to the presence of catalysts is accompanied by a corresponding change in the pre-exponential factor, A. This inter-dependence of activation energy E and pre-exponential factor A is called the compensation effect and frequently obeys an equation of the form:-

$$mE - \ln A = \text{constant} \tag{54}$$

where m is the proportionality constant and A, the pre-exponential factor, is indicative of the density of active sites on the carbon surface. The compensation effect operates as in Figure 10, where there is a cross-over through an iso-kinetic point of the Arrhenius plots of catalyzed and uncatalyzed reactions. An enhanced rate due to both a decrease in activation energy and increase in pre-exponential term is not reported. It is known for the activation energy to remain constant with an increase in the pre-exponential term.

11.2 Mechanism of catalysis

Two theories have been advanced to account for the effects of catalysis on carbon gasification reactions; the oxygen-transfer mechanism and the electron transfer-mechanism.

Oxygen-Transfer Mechanism

In oxygen-transfer theory, the catalyst is considered as an oxygen carrier that undergoes an oxidation-reduction cycle:-

$$MO + CO_2 \rightarrow MO.CO_2 \tag{55}$$

$$MO.CO_2 + C \rightarrow MO + 2CO \tag{56}$$

where MO represents a metallic oxide. The catalytic compound could include metals, M, as well as oxides, MO.

Electron-Transfer Mechanism

Long and Sykes (1950) applied electron theory to catalysis. Many of the known catalysts of gasification reactions either have unfilled electron shells and so can accept electrons from the carbon matrix or have labile electrons which can be donated to the carbon matrix. Transfer of electrons results in a redistribution of π-electrons in the carbon matrix. This weakens carbon-carbon bonds at edge sites and allows bond formation with an adsorbed oxygen atom which then requires less energy to be removed as a CO molecule, *e.g*:-

$$CO_3{}^{2-} + 2C_f \rightarrow 3CO + 2e^- \quad (57)$$

$$2M^+ + CO_2 + 2e^- \rightarrow M_2O + CO \quad (58)$$

$$M_2O + CO_2 \rightarrow M_2CO_3 \quad (59)$$

where e^- is an electron.

Today the oxygen-transfer mechanism is the most widely accepted of the two approaches. The catalysts are known to have localised behaviour i.e. the catalysed reaction only occurs at the point of contact of the catalyst and the carbon surface which is explained by the oxygen-transfer mechanism. Also, it is found (Amarigilio 1966) that the activation energy of catalysed oxidation of carbon is independent of the concentration of the catalyst present. This is not expected with the electron-transfer mechanism.

11.3 Understanding of catalysis by oxygen-transfer reactions

All of the gasification reactions of carbon can be catalyzed, mainly by Groups I and II and transition metals. The general understanding of the catalysis process is probably common to all reactions. However, as reaction temperatures and gas pressures differ significantly between the reactions, the energetics and concentrations of the important intermediate adsorbed surface species also differ, thus accounting for different reaction rates, activation energies and orders of reaction. The efficacy of an inorganic catalyst within a carbon, at least, is a function of the metal, the metal salt (or chemical state within the carbon), the state of distribution and degree of crystallinity within the carbon, concentration and access to the reacting gas (there may be others). Studies of different catalytic effects using different carbons and different catalysts for different gases at different temperatures with different methods of distribution of the catalyst in the carbon are not likely to give directly comparable results. Hence, the relative efficacies of catalysts, or pecking orders, differ throughout the literature. A useful review is that of Moulijn and Kapteijn (1986). The literature contains several detailed reaction schemes to explain catalysis by oxygen-transfer. Many are postulates because of the difficulty of obtaining analysis of reaction intermediates at reaction temperatures.

A recent, comprehensive study of the reaction of carbon dioxide with carbon with added catalyst is that of Cerfontain (1986). This study analyses, by FTIR spectroscopy, the intermediates in catalytic gasification by potassium carbonate (Cerfontain et al. 1983).

Upon heating a carbon/metal carbonate mixture to gasification temperatures

(>700°C) in an inert atmosphere, the salt decomposes to the oxide. This was observed by Cerfontain (1983) by using FTIR spectroscopy. The metal oxide may remain as the catalyst species, or, if unstable at gasification temperatures, may be reduced by the carbon to the metal:-

$$M_2CO_3 \rightarrow M_2O + CO_2$$

$$M_2O + C \rightarrow 2M + CO$$

The exact nature of the metal-carbon interaction is not clear. Various metal-carbon-oxygen complexes such as phenolates have been postulated (Cerfontain 1987) but this is an attempt to relate low temperature chemistry (<200°C) to high temperature conditions (>700°C). Most probably a non-stoichiometric catalyst-carbon structure exists at gasification temperatures, this being extremely sensitive to catalyst preparation and pre-treatments.

Introduction of carbon dioxide to the catalyst-carbon system at gasification temperatures results in an initial surge of carbon monoxide *e.g*:-

$$M + CO_2 \rightarrow M(O) + CO.$$

The amount of CO produced then decreases to a steady state value.

Cerfontain studied the composition of the catalyst using carbon-dioxide labelled with ^{18}O and ^{13}C. He showed that the catalyst contains chemisorbed carbon dioxide which rapidly exchanged with the carbon dioxide in the gas phase. This was further confirmed using a mixture of $^{13}CO_2/^{12}CO/He$ below gasification temperatures. It was found that only $^{12}CO_2$ was desorbed.

Rates of oxygen exchange between the catalyst and carbon dioxide were 10 times faster than rates of gasification. Amounts of chemisorbed carbon dioxide and bound oxygen were almost equal (Cerfontain et al. 1988). Rates of oxygen exchange were proportional to the CO partial pressure but independent of the CO_2 partial pressure.

These results imply that the formation of an oxygen-potassium-carbon complex is not the rate determining step and that the oxygen exchange reaction occurs at catalyst sites and not at carbon sites where it would involve C(O) complexes, the presence of which are dependent upon the partial pressures of both CO and CO_2. The rate controlling step is thus the desorption of C(O) as CO, the same as in the uncatalysed reaction. This leads to the conclusion that alkali catalysis of carbon gasification is due to an increase in the number of gasification sites compared with the uncatalysed reaction. Cerfontain et al. (1988) suggests that the chemisorbed oxygen is mobile. As oxygen is taken up by the catalyst then oxygen spillover (transfer) can occur from the catalyst to the carbon surface.

Hence, catalysis by inorganic metals or their oxides enhances the rate of removal of carbon which forms the carbon oxygen complex. Kinetic studies of different catalysts indicate changes in activation energies and pre-exponential functions. In studies of this complexity, there is no certainty that the mobile surface oxides are all identical, independent of reacting gas, temperature, pressure catalyst and catalyst preparation. Hence, differences in kinetic behaviour would not be unexpected.

Many transition metals e.g. iridium, rhodium, iron, cobalt, and nickel are also known to catalyse the reaction (McKee 1981; Baker et al. 1980). Copper catalyses the reaction by the reduction of CuO and Cu_2O to the metal at temperatures above 500°C (McKee 1970). Aluminium, zinc and tin are not catalysts as their oxides are stable in the presence of carbon.

11.4 Topography of catalytic gasification

Catalysis not only changes the kinetics of gasification reactions but also alters surface topography during gasification. Microscopy studies of the oxidation of graphites were made by Hennig (1966). McKee and Chatterji (1975), studying the oxidation of graphite flakes with sodium and rhubidium carbonates using a hot-stage microscope, observed that mobility of the catalyst particles was accompanied by the formation of a gasified channel on the graphite basal plane. They concluded that mobility and catalytic channelling were associated with the formation of a liquid peroxide.

Marsh et al. (1965), using transition electron microscopy, observed the topographical effects of oxygen and atomic oxygen on graphite and on graphite with iron. They observed the formation of hexagonal pits and hillocks on the graphite surface.

Uncatalysed gasification occurs at imperfections in the graphite basal plane e.g. vacancies, edges of basal planes and steps where one basal plane ends but a plane underneath continues. Catalysts are found to accumulate on a carbon surface at these sites due to the availability of electrons at imperfections in the lattice (Baker 1986). The following modes of surface gasification are attributed to carbon-catalyst interactions.

Pitting

It has been found that catalysts located at vacancies within a graphite basal plane form pits in the surface of the plane, Figure 11 (Jones 1988). The catalyst particle attacks perpendicular to the basal plane so forming a hexagonal hole. This can continue to increase in depth due to penetration of the catalyst and can also expand due to the edge recession of the hole. The larger the pit becomes, the more circular it appears. The formation of the pit is slower than the process of edge removal.

Edge Recession

Catalysts situated on graphite edge atoms and which have a very strong interaction with the carbon form a thin film of catalyst over the edge carbon atoms. Catalytic attack is then by edge recession (Figure 12) (Jones 1988).

Channelling

Catalyst particles at edges or steps of the graphite lattice act specifically on the edges and channel into the graphite layers. This mode of attack, observed by Baker (1986), occurs when the degree of interaction between the catalyst and the carbon i.e. the degree of wetting, is less than that observed when edge recession occurs. During the formation of a channel, the catalyst particle appears to be fluid. In some cases, as the channel proceeds, particles of catalyst are left behind on the channel walls. This allows the walls to gasify further and the width of the channel to increase, whereas the catalyst particle itself becomes smaller and the progressing channel becomes narrower. The result is a channel with a fluted appearance (Figure 13a) (Jones 1988). The channel can be straight or can change direction due to structural changes in the carbon being gasified (Figure 13b).

Controlled Atmosphere Electron Microscopy (CAEM) has been used to study the topographical action of a catalyst on graphite during gasification. Baker (1986) has undertaken several studies using CAEM to observe the electrolytic behaviour of metals with different oxidising gases.

CAEM allows a sample to be studied in situ. The specimen, a single graphite crystal 15-100 nm thick, is surrounded by reactant gas at high pressure, while the rest of the microscope remains evacuated. The catalyst particle appears to eat its way through the carbon structure by pitting, edge recession or channelling. Thus, the catalyst acts at its point of contact with the carbon which complies with the theories that a metal-carbon interaction occurs (Cerfontain 1986). Rodriguez et al. (1987) used CAEM to study the mode of catalytic attack for carbons with different

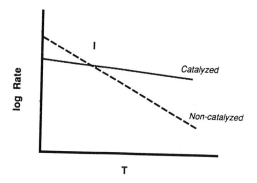

Figure 10. Arrhenius plots of catalyzed and non catalyzed gasification showing an iso-kinetic point **I**.

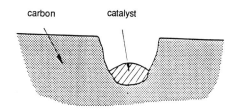

Figure 11. Pitting of a graphite basal plane.

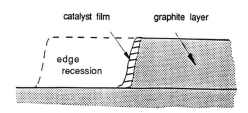

Figure 12. Edge recession of graphite basal plane due to the presence of a catalyst film.

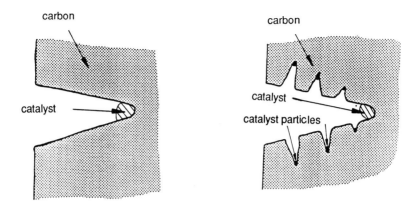

Figures 13 (a) and (b). Channeling action of a catalyst resulting in fluted channels

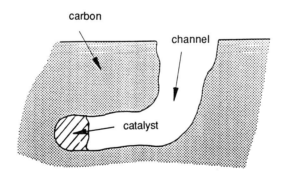

Figure 13 (c). Changing direction of a channeling catalyst particle.

optical textures. It was found that:

i) isotropic carbons or carbons with very limited anisotropy formed shallow pits at the low temperature of 375°C.
ii) anisotropic carbons formed deep pits at high (>525°C) temperatures and edge recession occurred especially in areas with highly ordered structures.

Recently, using thermodynamics, Rodriguez et al. (1987) and Baker (1986) have explained that metals, which adsorb oxygen dissociatively and readily form oxides, form strong interfacial bonds with carbons and so result in edge recession. Metals which adsorb oxygen non-dissociatively and tend to remain in the noble state do not exhibit a strong interaction with oxygenated carbons and exist as discrete particles at graphite edges which they attack by the channelling mode.

1 2 INHIBITION OF THE GAS-CARBON REACTION

The control or reduction of reactivity of carbonaceous materials can be attempted by five main routes. Two of these are involved with the structure of the solid carbon:-

a) The reduction of open porosity so restricting the access of gas to the carbon surface.

b) The improvement in the crystalline order within the carbon matrix which reduces the number of sites at which reaction can occur most easily.

The other three methods are:-

c) The exclusion of catalysts which enhance the gasification of carbon.

d) The introduction of atoms or groups which reduce gasification of carbon (i.e. the "negative catalysts").

e) The formation of a glassy layer on carbon surfaces so forming a barrier to the gas.

Although both of the methods d) and e) have been called inhibition of the gas-carbon reaction it is only d) which is true inhibition (McKee 1981). This inhibition is understood as a blocking of the possible active sites on the carbon surface (Allardice and Walker 1970).

A range of substances, mainly based on phosphorous, halogen and boron compounds, have been identified as acting as inhibitors to oxidizing gases (McKee et al. 1984a, 1984b and McKee and Spiro 1985). They include organophosphorous compounds, phosphates, phosphonyl chloride, sulphonyl chloride, halogen gases, organohalogens, boron, boron oxide, organo-boron compounds and some transition metals.

The mode of action of inhibitors has been widely investigated. In some cases, the inhibitor reduces the effect of intrinsic catalysts, e.g. deactivating them by forming stable phosphates (Magne et al. 1971), but the principal effect is to adsorb stable complexes on to the active sites in the carbon structure. The most effective inhibitor known is phosphorous oxychloride ($POCl_3$) which has been shown to reduce the oxidation of nuclear graphite by a factor >10 (Hawtin and Gibson 1970).

Possible active sites include point defects and screw dislocations within the graphitic material and, of course, the large variety of surface structures forming the less-ordered parts of the material. All of these can be protected to some extent by the inhibitors mentioned above but increasing structural order by graphitization is usually more effective. However, at the limit of structural order within carbon materials there are two types of active site to be considered, namely the zig-zag and armchair edges of the graphite lamellae (see Figure 14).

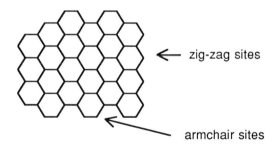

Figure 14 Active sites at edges of graphite lamellae

The relative reactivity of these sites is difficult to assess and may vary from reactant to reactant depending, for example, on reaction temperature and steric effects. The oxidation of unprotected graphite by atomic oxygen or water vapour leads to hexagonal pits in the structure developing from point defects. Both zig-zag {1011} (Yang and Duan, 1985) and armchair {1121} (McKee and Spiro 1985) faces have been reported as making up the edges of these pits. Oxidation by molecular oxygen or carbon dioxide usually results in a round pit with edges, therefore, exhibiting a range of sites. Zig- zag edges are interpreted as being caused by preferential attack at the armchair sites (Thomas 1965 and Yang and

Duan 1985). Figure 15 illustrates the process and recent computer simulation work has provided a quantitative justification for this interpretation (Brown 1987). In contrast, the armchair edges suggest that it is the zig-zag sites which are most reactive (McKee and Spiro 1985). This view is supported by Stein and Brown (1985) who concluded, from structure-resonance theory calculations, that the zig-zag sites were more reactive than the armchair sites.

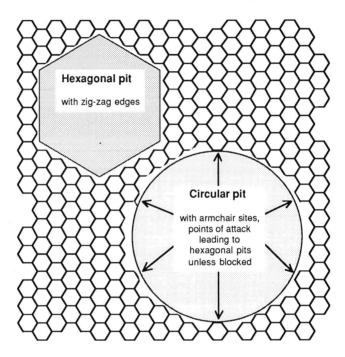

Figure 15 Pits formed in graphitic layers

Whatever the correct mechanism may be, blocking of reactive sites with chlorine, or other complexes, results in round pits. Hippo et al. (1986) observed this and interpreted the blocking to be of the zig-zag sites. They also showed that the treatment of both polycrystalline graphite and carbon-carbon composites with phosphoric acid is effective in inhibiting gasification. Activation energy for the reaction is increased and the change from chemical to diffusional (porosity) control is moved to higher temperatures.

The identification of the sites at which inhibitors act is problematical as it is probably necessary to block the alternative sites before any inhibiting effect can be detected without ambiguity. The search for effective methods of protecting carbon materials from high temperature gasification continues.

References

Ahmed, S. and Back, M.H. (1985). The role of the surface complex in the kinetics of the reaction of oxygen with carbon. Carbon **23**, 513.

Allardice, D. J., and Walker, P.L. Jr., (1970). The effect of substitutional boron on the kinetics of the carbon-oxygen reaction. Carbon **8**, 375.

Amarigilio, H. and Duval, X. (1966). Etude de la combustion catalytique du graphite. Carbon **4**, 323.

Arthur, J. (1956). Reactions between carbon and oxygen. Trans. Faraday Soc. **47**, 164.

Baker, R.T.K. and Sherwood, R.D. (1980). Catalytic oxidation of graphite by iridium and rhodium. J. Catalysis **61**, 378.

Baker, R.T.K. (1986). Metal catalysed gasification of graphites. 'Carbon and Coal Gasification', Eds. J. Figueiredo, J.A. Moulijn, Proc. NATO Advanced Science Institute, Alvar, Portugal, pp. 231-268.

Baker, R.T.K. (1986). Factors controlling the mode by which a catalyst operates in the graphite-oxygen reaction. Carbon **24**, 715.

Barton, S.S. and Harrison, B.H. (1975). Acidic surface oxidation structures on carbon and Graphite. Carbon **13**, 283.

Bedjai, G., Orbach, N.N. and Riesenfeld, F.C. (1958) Reaction of nitric oxide with activated carbon and hydrogen. Ind. Eng. Chem. **50**, 1165.

Blackwood, J.D. and Taggart, F.K. (1959). Reactions with carbon with atomic gases. Aust. J. Chem. **12**, 533.

Blake, J.H., Bopp, G.R., Jones, J.F., Miller, M.G. and Tambo, W. (1967). Aspects of the reactivity of porous carbons with carbon dioxide. Fuel **46**, 115.

Board, J.A. (1966). The thermal oxidation of nuclear graphite in carbon dioxide. Proc. 2nd Conf. on Ind. Carbon and Graphite, S.C.I., London, pp. 277-289.

Brown, E. F., (1987). The equilateral nature of the hexagonal etch pit developed during carbon oxidation. Carbon **25**, 617

Cerfontain, M.B. and Moulijn, J.A. (1983), Alkali catalysed gasification reactions studied by in situ FTIR spectroscopy. Fuel **62**, 256.

Cerfontain, M.B. (1986), Alkali catalysed carbon gasification, Ph.D. Thesis, University of Amsterdam, The Netherlands.

Cerfontain, M.B., Agalianos, D. and Moulijn, J.A. (1987). CO_2 step-response experiments during alkali catalysed carbon gasification; evaluation of the so-called CO overshoot. Carbon **25**, 351.

Cerfontain, M.B., Kapteijn, F. and Moulijn, J.A. (1988). Characterization of alkali carbonate catalysts for carbon gasification with ^{18}O labelled CO_2. Carbon **26**, 41.

Day, R.J., Walker Jr., P.L. and Wright, C.C. (1958). The carbon-oxygen reaction at high temperatures and high gas flow rates. Ind. Carbon Graphite Conf., SCI, London, pp. 348-370.

Ergun, S. and Mentser, M. (1965), Chemistry and Physics of Carbon, Ed. P.L. Walker Jr., Marcel Dekker, Vol. **1**, 203.

Ergun, S. (1956) Kinetics of the reaction of carbon dioxide with carbon. J. Phys. Chem. **60**, 480.

Feistel, P.P., van Heek, K.H., Juntgen, H. and Pulsifer, A.H. (1977). Gasification of a German bituminous coal with H_2O, H_2 and H_2O/H_2 mixtures. Am. Chem. Soc. Div. Fuel Chem. Preprints, **22**, (1), 53.

Fischbach, D.B. (1971). Chemistry and Physics of Carbon, Ed. P.L. Walker Jr., Marcel Dekker, New York, Vol. **7**, 1.

Frank-Kamenetskii, D.A. (1969), Diffusion and Heat Transfer in Chemical Kinetics, Plenum Press, New York.

Franklin, R. (1949). A study of the fine structure of carbonaceous solids by measurements of true and apparent densities. Trans. Faraday Soc. **45**, 668

von Fredersdorff, C.G. and Elliot, M.A. (1963). Coal Gasification, Chemistry of Coal Utilization, Ed. H.H. Lowry, Supplementary Volume, Wiley, New York, 892.

Gadsby, J., Long, F.J., Sleightholm, P. and Sykes, K.W. (1948). The mechanism of the carbon dioxide-carbon reaction. Proc. Roy. Soc. A193 357.

Garcia, X. and Radovic, LR. (1986). Gasification reactivity of Chilean coals. Fuel **65**, 292.

Hawtin, P. and Gibson, J. A., (1970). Inhibition of air oxidation of large specimens of graphite with phosphorus oxychloride. Proc. 3rd Conf. Industrial Carbon and Graphite, S.C.I., London 1970, p 309.

Hayward, D.O. and Trapnell, B.M.W. (1964), Chemisorption, Butterworths, London.

Heilbronner, E. and Bock, H. (1976), The HMO model and its application. Wiley-Interscience, New York.

Hennig, G.R. (1966), Chemistry and Physics of Carbon, Ed. P.L. Walker Jr., Marcel Dekker, New York, Vol. **2**, 1.

Hippo, E. and Walker Jr., P.L. (1975). Reactivity of heat-treated coals in carbon dioxide at $900^\circ C$. Fuel **54**, 245.

Hippo, E.J., Murdie, N., Kowbel, W. and Wapner, P. A., (1986). NASA Conference Publication 2482, Proceedings of a Joint NASA/DOD Conference, Florida, USA, p.457.

Jenkins, R.G., Nandi, S.P. and Walker Jr., P.L. (1973). Reactivity of heat-treated coals in air at $500^\circ C$. Fuel **52**, 288.

Jenkins, R.G. and Piotrowski, A. (1978). Role of carbon active sites in the oxidation of coal char. 194th National Meeting ACS, New Orleans, Louisiana, USA, Aug. 31 - Sept. 4, 32, **4**, 147.

Jones, L.A. (1988). Catalysis of carbon by sodium. Ph.D.Thesis, University of Newcastle upon Tyne, U.K.

Khan, M.R. (1987). Significance of char active surface area for appraising the reactivity of low - and high - temperature chars. Fuel, **66**, pp. 1626-1634.

Kochling, K.H., McEnaney, B., Rozploch, F. and Fitzer, E. (1982), "International Committee for Characterization and Terminology of Carbon ", First Publication of 30 Tentative Definitions, Carbon **20**, 445.

Koenig, P.C., Squires, R.G. and Laurendeau, N.M. (1985). Evidence for two-site model of char gasification by carbon dioxide. Carbon **23**, 531.

Krevelen, D.W. van (1961), Coal, Elsevier, New York.

Laine, N.R., Vastola, F.J. and Walker Jr., P.L. (1963a). The importance of active surface area in the carbon-oxygen reaction. J. Phys. Chem. 67, 2030.

Laine, N.R., Vastola, F.J. and Walker Jr., P.L. (1963b). The role of the surface complex in the carbon-oxygen reaction. Proceedings of the Fifth Conf. Carbon, Vol. 2, Pergamon Press, New York, pp. 211-219.

Lang, F.M., Magnier, P. and May, S. (1962). Etude de l'oxydation des carbones par l'air et l'anhydride carbonique en fonction de leur texture et de leur etat de purete. Proc. 5th Conference on Carbon, Pennsylvania State University 1961, Pergamon Press, Oxford, Vol. **1**, pp. 171-193.

Lewis, I.C. and Edstrom, T. (1963). Thermal reactivity of polynuclear aromatic hydrocarbons. J. Org. Chem. **28**, 2050.

Lewis, J.B. (1970), Modern Aspects of Graphite Technology, Ed. L.C.F. Blackman, Ch. 4, Acad. Press, London, pp. 129-199.

Long, F.J. and Sykes, K.W. (1950). The catalysis of the oxidation of carbon. J. Chim. Phys. **47**, 361.

McKee, D.W. (1970). The copper-catalysed oxidation of graphite. Carbon **8**, 131.

McKee, D.W. and Chatterji, D. (1975). The catalytic behaviour of alkali metal carbonates and oxides in graphite oxidation reactions. Carbon **13**, 381.

McKee, D.W. (1981), Chemistry and Physics of Carbon, Eds. P.L. Walker Jr. and P.A. Thrower, Marcel Dekker, New York, Vol. **16**, 1.

McKee, D.W. (1983). Mechanisms of the alkali metal catalysed gasification of carbon. Fuel **62**, 170.

McKee, D. W., Spiro, C. L. and Lamby, E. J., (1984a). The inhibition of graphite oxidation by phosphorus additives. Carbon **22**, 285.

McKee, D. W., Spiro, C. L. and Lamby, E. J., (1984b). The effects of boron additives on the oxidation behaviour of carbons. Carbon **22**, 507.

McKee, D. W. and Spiro, C. L., (1985). The effect of chlorine pretreatment on the reactivity of graphite in air. Carbon **23**, 427.

Madison, J.J. and Roberts, R.M. (1963). Pyrolysis of aromatics and related heterocyclics. Ind. Eng. Chem. **50**, 237.

Madley, D.G. and Strickland-Constable, R.F. (1953). The kinetics of the oxidation of charcoal with nitrous oxide. Trans. Faraday Soc. **49**, 1312.

Magne, P., Amariglio, H. and Duval, X., (1971).Kinetic study of graphite oxidation inhibited by phosphates. Bull. Chim. Fr. A6, 2005.

Marsh, H., O'Hair, T.E. and Reed, R. (1965). Oxidation of carbon and graphites by atomic oxygen. An electron microscope study of surface changes. Trans. Faraday Soc. **61**, 285.

Marsh, H. (1979). A review of the growth and coalescence of mesophase (nematic liquid crystals) to form anisotropic carbon, and its relevance to coking and graphitization. Proc. 4th London International Carbon and Graphite Conference, Sept. 1974. Society of Chemical Industry, London, pp. 2-38.

Marsh, H., Taylor, D.A. and Lander, J.R. (1981). Kinetic study of gasification by oxygen and carbon dioxide of pure and doped graphitizable carbons of increasing heat-treatment temperature. Carbon **19**, 375.

Morgan, W.C. and Thomas, M.T. (1982). The inverse oxidation phenomenon, Carbon **20**, 71.

Moulijn, J.A. and Kapteijn, F. (1986). Catalytic gasification. 'Carbon and Coal Gasification', Eds. J. Figueredo, J.A. Moulijn, (1985), Proc. NATO Advanced Science Institute, Alvar, Portugal, pp. 181-195.

Nahas, N.C. (1983). Exxon catalytic coal gasification process. Fundamentals to flowsheets. Fuel **62**, 239.

Phillips, R., Vastola, F.J. and Walker Jr., P.L. (1970). The thermal decomposition of surface oxides formed on graphon. Carbon **8**, 197.

Pinnick, H.T. (1956), Electronic properties of carbons and graphites. Proc. 1st and 2nd Conferences on Carbon, Waverley Press Inc., University of Buffalo, US, 1953 and 1955, USA, pp. 3-11.

Radovic, L.R., Walker Jr., P.L. and Jenkins, R.G. (1983). Effect of lignite pyrolysis conditions on calcium oxide dispersion and subsequent char activity, Fuel **62**, 209.

Rakszawski, J.F., Rusinko, F. Jr. and Walker Jr., P.L. (1963). Catalysis of the carbon-carbon dioxide reaction by iron. Proceedings of The Fifth Carbon Conference, Vol. **2**, pp. 243-251.

Reif, A.E. (1952). The mechanism of the carbon dioxide-carbon reaction. J. Phys. Chem. **56**, 785.

Rodriguez, N.M., Marsh, H., Heintz, E.A., Sherwood, R.D. and Baker, R.T.K. (1987), Oxidation studies of various petroleum cokes. Carbon **25**, 629.

Rodriguez-Reinoso, F., Thrower, P.A. and Walker Jr., P.L. (1974). Kinetic studies of the oxidation of highly oriented pyrolytic graphites. Carbon **12**, 63.

Rossberg, M. (1956). Experimentelle Ergebnisse uber die Primarreaktionen bei der Kohlenstoffverbrennung. Z. Electrochem. **60**, 952.

Smith, W.R. and Polley, M.H. (1956). The oxidation of graphitised carbon black. J. Phys. Chem. **60**, 689.

Smith, I.W. (1978). The intrinsic reactivity of carbons to oxygen. Fuel **57**, 409.

Smith, R.N., Lesnini, D., Mooi, J. (1957). The oxidation of carbon by nitrous oxide. J. Phys. Chem. **61**, 81.

Stein, S.E. and Brown, R.L. (1985). Chemical theory of graphite-like molecules. Carbon **23**, 105.

Strange, J.F. and Walker Jr., P.L. (1976). Car;bon-carbon dioxide: Langmuir-Hinshelwood kinetics at intermediate pressures. Carbon **14**, 345.

Swinborne-Sheldrake, R. and Herndon, W.C. (1975). Kekule structures and resonance energies of benzenoid hydrocarbons. Tetra. Lett. **10**, 755.

Thomas, J.M. (1965). Chemistry and Physics of Carbon, Microscopy studies of graphite oxidation. Ed. P.L. Walker, Jr., Marcel Dekker, New York, Vol. **1**, 121.

Thomas, J.M. and Thomas, W.J. (1967). Introduction to the Principles of Heterogeneous Catalysis, Academic Press, London.

Thomas, J.M. (1970). Reactivity of carbon: some current problems and trends. Carbon **8**, 413.

Tomita, A., Mahajan, O.P. and Walker Jr., P.L. (1977). Catalysis of char gasification by minerals. Am. Chem. Soc. Div. Fuel Chem. Preprints **22**, 4.

Vastola, F.J., Hart, P.J. and Walker Jr., P.L. (1964). A study of carbon-oxygen surface complexes using O^{18} as a tracer. Carbon **2**, 65.

Walker Jr., P.L., Rusinko Jr., F. and Austin, L.G. (1959). Gas reactions of carbon. Advan. Catalysis **11**, 133.

Walker Jr., P.L., Shelef, M. and Anderson, R.A. (1968). Chemistry and Physics of Carbon, Ed. P.L. Walker Jr., Marcel Dekker, New York, Vol. **4**. 287.

Watts, H. (1958). The oxidation of charcoal by nitric oxide and the effect of some additives. Trans. Faraday Soc. **54**, 93.

Wicke, E. (1955). Contributions to the combustion mechanism of carbon. Fifth Symp. Combustion, Reinhold, New York, 245.

Yang, R. T. and Duan, R. Z., (1985) Kinetics and mechanism of gas-carbon reaction: conformation of etch-pits, hydrogen inhibition and anisotropy in reactivity. Carbon **23**, 325.

Zielke, C.W. and Gorin, E. (1955). Kinetics of carbon gasification. Ind. Eng. Chem. **47**, 820.

Chapter 5

Porosity in Carbons and Graphites

B. McEnaney and T.J. Mays

School of Materials Science, University of Bath, Bath, BA2 7AY, U.K.

Summary.

In this Chapter various types of porosity in carbons and graphites are described together with their effects on some important properties, e.g. density and strength. Various methods for characterising micropores, mesopores and macropores are reviewed critically. Surface areas can be estimated from the Brunauer-Emmett-Teller (BET) equation applied to gas adsorption, or from the methods of Debye, Porod and Guinier applied to small angle X-ray scattering, SAXS. Gas adsorption and SAXS can also be used to estimate the fractal dimensions of pore surfaces. The BET equation is severely limited when applied to micropores, because micropore sizes are commensurate with the dimensions of adsorbate molecules. Alternatively, micropore volumes can be obtained from the Dubinin-Radushkevich equation applied to gas adsorption. Mesopore volumes and sizes can be obtained from capillary condensation theory applied to adsorption hysteresis, although the method is subject to limitations, particularly as a result of neglecting of pore interconnections. Mercury porosimetry is used to obtain macropore volumes and sizes, but also neglects pore interconnections. Alternative estimates of macropore sizes are available from measurements of gaseous permeation and diffusion and mercury permeametry. Image analysis can provide measurements such as macropore shape factors and spatial distributions, which cannot be obtained by other methods, but a limitation is that the measurements are obtained from a two-dimensional image of a three dimensional object.

POROSITY IN CARBONS AND GRAPHITES

Brian McEnaney and Timothy J. Mays

School of Materials Science, University of Bath, Bath BA2 7AY, UK

1 INTRODUCTION

The great majority of manufactured carbons and graphites contain voids or pores which either result from the manufacturing process or are inherent in the microstructure of the raw materials. After manufacture, initial voids may be developed or new voids created as a result of mechanical, chemical or heat treatments, or exposure to nuclear radiations. In some cases, e.g., active carbons, porosity is desirable and may even be enhanced to facilitate adsorption of fluids. In others, e.g., nuclear graphites, extensive porosity is undesirable, since it allows transport of oxidizing gases to within the graphite structure, leading to corrosion and the subsequent deterioration in properties such as mechanical strength. In this Chapter various types of voids and porosity in carbons and graphites are described together with their effects on some important properties of these materials, and different methods for characterising porosity in carbons and graphites are reviewed critically.

1.1 Classifications of porosity

It may be useful to begin a classification of porosity by defining the terms 'void' and 'pore'. A void is an empty space which in solid materials occurs where there is a discontinuity in the array of atoms and molecules, i.e., where their electron density falls to zero. Small angle X-ray scattering due to the electron density transition at the solid-void interface may be used to obtain parameters of the voidage (see the main section on small angle X-ray scattering). The word 'pore' comes from the Greek word $\pi o \rho o \sigma$, meaning a passage. In this sense a pore is a class of void which is connected to the external surface of a solid and will allow the passage of fluids into, out of, or through a material. In the scientific literature on porous solids the terms 'open pore' and 'closed pore' are used, the former denoting a pore which is connected to the external surface of the solid and the latter denoting a void which is not so connected. To conform with general usage, the terms 'open pore' and 'closed pore' will be employed in this Chapter, although 'void' is perhaps a better general term. The total pore volume fraction, V_T, is given by $V_T = v_p / (v_p + v_s)$, where v_p and v_s are the specific volume of pores and solid carbon respectively, and the solid volume fraction $V_S = 1 - V_T$. Also, $V_T = V_O + V_C$, where V_O and V_C are the open and closed pore volume fractions respectively.

Porous materials may be consolidated or unconsolidated. A bed of activated carbon granules is an example of an unconsolidated porous medium. The pore structure comprises the intraparticulate voidage, i.e., the inherent porosity of the particles, and the interparticulate voids, i.e., the spaces between the granules. In contrast, in consolidated materials, e.g., engineering graphites, the solid is a continuous structure within which there is a network of pores of different sizes and shapes, as well as isolated pores. When fluid transport through pores is considered (see the sub-section on fluid transport in pores), the open pore volume, V_O, may be sub-divided into the transport pore volume, V_t, and the blind pore volume V_b. Transport pores are those pores in which a concentration gradient exists during steady state or time-independent fluid flow through the material. Blind pores are connected to transport pores by a single opening so that in them concentration gradients and hence fluid flow only occur during unsteady state or time-dependent flow. Examples of the different pore types are shown schematically in Figure 1.

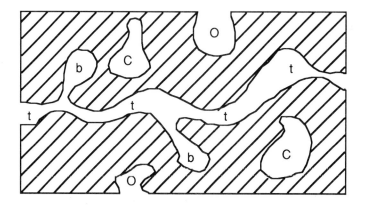

Figure 1. Different types of porosity in a porous solid.
O - open pores; C - closed pores; t - transport pores; b - blind pores.

Pores in carbons and graphites may also be classified by their size and shape. The spectrum of pore sizes extends from molecular dimensions to massive defects many centimetres in size in large carbon artefacts. A classification based upon pore size was proposed by The International Union of Pure and Applied Chemistry, IUPAC, (Sing et al.,1985) as follows:-

        micropores   -   width less than 2 nm;
        mesopores   -   width between 2 and 50 nm;
        macropores   -   width greater than 50 nm.

This classification, which is an arbitrary one and based upon the adsorption characteristics of porous solids, is now widely adopted, even when adsorption is not being considered, and will be used in this Chapter.

The shapes of pores in carbons and graphites vary from slit-shaped cracks to spheroidal bubbles. Cracks may follow tortuous paths through the solid and may be connected to other pores to form an extensive and irregular network. The shapes of pores can have important effects on some properties of carbons and graphites, e.g., mechanical strength and thermal expansivity, but the characterisation and classification of pores by shape is much less advanced than classification by size. Also network effects arising from the connectivity of pores are important in some situations, e.g., the interpretation of adsorption hysteresis and percolation of fluids, but are often neglected in standard treatments of porosity (see the section on mesoporous carbons). Some examples of the numerous and diverse types of pores in different carbons and graphites are given in the next sub-section.

1.2 Some examples of porosity in carbons and graphites

Figure 2a is an optical micrograph of pores in an active carbon made from almond shells. The sections of the pores shown in the micrograph are approximately elliptical with a mean equivalent circle pore diameter (see the sub-section on image analysis) of about 15 µm, so that they are large macropores. They are relics of the cellular structure of the precursor material. This is a general characteristic of macropores in active carbons derived from lignocellulosic precursors. On the atomic scale, activated carbons have a very disordered structure, as indicated in the high resolution electron micrograph of a cellulose carbon, Figure 2b. Electron microscopical studies have led to models for the ultrastructure of activated carbons consisting of a twisted network of defective carbon layer planes cross-linked by aliphatic bridging groups. Micropores are formed in the interlayer spacings with widths in the range 0.34 - 0.8 nm. It is the micropores in activated carbons which have the greatest influence upon gas adsorption, while macropores and mesopores are important in transport of fluids to and from the micropores (see the main section on microporous carbons).

Macropores in glassy carbons are illustrated in Figure 3 which shows spherical pores, mean diameter about 4 µm, in a glassy carbon formed from a liquid resole phenolic resin precursor. These are closed macropores formed by bubbles of gases released during curing of the resin which are trapped in the increasingly viscous resin matrix. The bubble structure is preserved during carbonisation. Resin carbons also contain micropores which are similar to those in activated

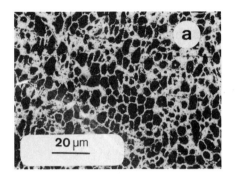 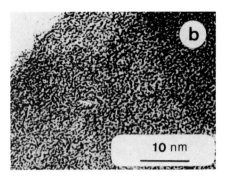

Figure 2. a - Cellular macropores in an almond shell carbon; b - HREM of a cellulose carbon.

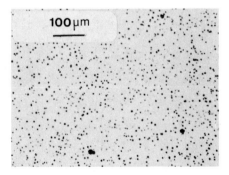

Figure 3. Macropore structure of a glassy carbon showing spherical bubble pores.

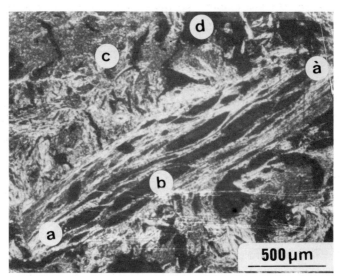

Figure 4. Macropores in an electrode graphite.
a-à - needle-coke filler particle with volumetric shrinkage cracks, b;
c - binder phase; d - gas evolution pores in the binder phase.

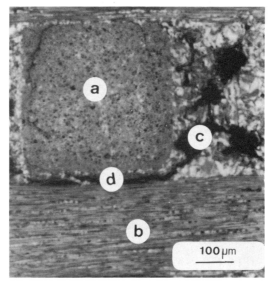

Figure 5. Macropore structure of a 3-D carbon-carbon composite.
a, b - mutually perpendicular fibre bundles; c - gas evolution pores in pitch pocket;
d - bundle-bundle interface cracks

carbons, Figure 2b. However, because resin carbons are formed from highly cross-linked precursors and are not activated, the majority of micropores in them are usually closed.

Several types of pores and cracks in an engineering graphite are shown in Figure 4; each of them is formed at a different stage in the fabrication process. Features marked 'b' are volumetric shrinkage cracks formed during carbonisation and calcination of the needle coke particles, 'aà'. The cracks have widths up to about 100 μm and are up to about 1 mm long. Graphitic basal planes lie parallel to these cracks so that there is good thermal and electrical conductivity in the direction parallel to the shrinkage cracks, i.e., along the long axis of the filler particles. The superior electrical conductivity along the long axis of extruded graphite electrodes results from the orientation of the needle coke particles during the extrusion stage of manufacture. The shrinkage cracks in filler particles also accommodate the anisotropic thermal expansion of the graphite during heating and so contribute to the low bulk thermal expansion of graphite electrodes.

The features marked 'd' in Figure 4 are globular macropores in the binder phase, 'c'. These are formed by evolution of gases during the baking stage of manufacture when the pitch binder is converted to carbon. The globular pores in the binder phase are often linked together by a network of fine shrinkage cracks. These cracks are formed as a result of shrinkage of the binder phase during baking, and also by anisotropic contraction on cooling from graphitisation temperatures (Mrozowski, 1956). Mrozowski cracks may be formed by extension of the shrinkage cracks so that it is difficult to distinguish between them. The macroporous networks in the binder phase are open, whereas the shrinkage cracks in the filler particles are usually closed, probably because they are sealed with binder pitch during the mixing stage of manufacture. The shrinkage cracks in the filler particles may be opened up by oxidation by gases. Figure 4 illustrates how a carbon or graphite can have different types of macropores and cracks, having different sizes and shapes and effecting the properties of the material in different ways. Methods for characterizing porosity which take no account of this diversity, e.g., mercury porosimetry, can sometimes be misleading (see the main section on macroporous carbons).

Figure 5 shows a system of pores and cracks in a 3-D carbon-carbon composite. The composite is based upon a 3-D orthogonal weave of fibre bundles which have been impregnated with pitch. The features marked 'a' and 'b' are mutually perpendicular fibre bundles. A few globular gas evolution pores, 'c', similar to those in the engineering graphite, Figure 4, can be seen in the pitch 'pocket'. These pores are formed during baking stages of manufacture. The globular pores are linked to cracks in the pitch carbon, which in turn are linked to cracks at the bundle-bundle interfaces, 'd'. The latter cracks form as a result of complex stresses generated during the fabrication of these materials. The presence of these pores and cracks has an important bearing upon the thermomechanical properties of these composites, but a full, quantitative understanding of these factors has not been achieved.

2 EFFECTS OF POROSITY ON PROPERTIES OF CARBONS

The presence of pores in a material will have significant effects upon many technically-important properties and most obviously upon density and upon surface area; these properties will be discussed in more detail in the next two main sections. The presence of porosity in carbons and graphites adversely affects mechanical properties such as strength and elastic modulus since it reduces the volume of solid within which the stresses are distributed . Also, pores can act as sites for local stress concentration which can initiate failure. Pickup et al. (1986) showed that strength, modulus and fracture mechanics parameters, such as critical stress intensity factor decreased with increasing porosity. It was also demonstrated that high aspect ratio pores, e.g., thermal strain cracks, shrinkage cracks and pores elongated as a result of thermal oxidation, reduced mechanical properties to a greater extent than more equiaxed pores, confirming expectations from theoretical fracture mechanics.

Electrical and thermal conductivities decrease with increasing porosity as a result of the reduced amount of solid to conduct electrons and heat. Thermal expansivity in porous graphites is considerably reduced compared to the value for single crystal graphites, due to the accommodation of thermal expansion within pores, particularly within shrinkage cracks and thermal strain cracks lying parallel to basal planes, Figure 4.

Pores also have an important influence upon gasification reactions, at temperatures where diffusion of reactant and product gases in the pores influences the rate. Clearly, gasification is influenced only by open porosity, i.e., pores accessible to the reactant gas, whereas mechanical, thermal and electrical properties are influenced by the total open and closed porosity. The temperature range where in-pore diffusion influences the rate is denoted as Zone II in the reaction scheme of Hedden and Wicke (see Walker et al., 1959). Zone I occurs at lower temperatures when the chemical reaction at the carbon surface controls the rate, and Zone III occurs at higher temperatures when the gasification rate is controlled by diffusion of reactant and product gases through an external gaseous boundary layer. In Zone I the gasification rate is directly proportional to the surface area of the carbon and therefore is related to the total open porosity, but it is not dependent upon pore size. (Strictly, gasification rate is directly proportional to active surface area, i.e., that portion of the total surface area participating in the gasification reaction.) In Zone II the gasification rate is dependent upon both the (active) surface area and the diffusivity of reactant and product gases in the pores. In turn, diffusivity in pores, i.e., the effective diffusivity, is related to the open pore structure, i.e., to the distribution of pore sizes and their connectivity. In Zone III the gasification rate is directly proportional to the external (active) surface area and independent of the porous texture of the carbon. Gasification of carbons and graphites in industrial situations is frequently influenced by in-pore diffusion. Examples include coal char gasification in steam and CO_2, gasification of graphites in nuclear reactors, and gasification of carbon anodes used in aluminium smelters.

3 THE DENSITIES OF CARBONS

The density of an engineering graphite can most easily be obtained by measuring the weight and dimensions of a specimen of regular shape, e.g., a rectangular block; such a method is described in ASTM (C 559)-79. The measured block density, ρ_b, includes the volumes of solid carbon and of open and closed pores; the total pore volume, V_T, can be obtained from the formula:-

$$V_T = 1 - \rho_b / \rho_t \qquad (1)$$

where ρ_t is the density of the solid carbon which is usually assumed to be equal to the X-ray density of single crystal graphite (2.267 g cm^{-3}). The contributions to V_T from open and closed pores, V_O and V_C respectively, can be estimated if the density of the graphite is determined in helium, ρ_{He}, a fluid which is presumed to penetrate the entire open pore volume, without being adsorbed on the surface of pore walls. The resulting equation is:-

$$V_C = 1 - \rho_{He} / \rho_t \qquad (2)$$

where $V_O = V_T - V_C$.

For disordered, porous systems, such as activated carbons, the relationships between density and porosity are not so straightforward. These materials are usually manufactured in particulate or granular form and a density of practical interest is the apparent or bulk density, ρ_a, which is the density of the material including the interparticulate voids; a method for determining ρ_a for activated carbons is in ASTM (D 2854)-30. The particle density of activated carbons is analogous to the block density of engineering graphites, in that it includes the volumes of the solid carbon and of open and closed pores. The particle density can only be measured using a dilatometric fluid which fills the interparticulate voids without penetrating the intraparticulate voids. In practice it is difficult to ensure that this condition is fulfilled, although a method based upon mercury densities is available [ASTM (C699)-45]. The solid phase in activated carbon is certainly not graphitic so that $\rho_t = 2.267$ g cm^{-3} cannot be assumed. Helium densities have been used to estimate the density of solid carbon in activated carbons, but, if there is significant closed porosity, then the ρ_t will be underestimated. Also, there is evidence that helium is adsorbed in micropores in some activated carbons (Sutherland, 1967) leading to an overestimate of ρ_t. The foregoing discussion shows that reliable estimates of pore volumes for activated carbons and other disordered, particulate carbons from measurements of densities can be fraught with difficulties and uncertainties.

Measurement of the densities of porous carbons in a range of fluids of different molecular size which penetrate the pore space to different extents can be used to construct micropore size distributions. The method is laborious and has its own uncertainties, e.g., the selection of the critical dimension of the dilatometric molecule to determine pore size, and problems due to density drift caused by slow penetration of the pore space; this method has been reviewed in detail by Spencer (1967).

4   SURFACE AREAS FROM GAS ADSORPTION

The specific surface area (m^2 g^{-1}) of porous carbons and graphites is most usually determined from gas adsorption measurements using the Brunauer-Emmett-Teller (BET) theory (see Gregg and Sing, 1982). It should be noted that this method gives the surface area of open pores. Nitrogen at its boiling point of 77 K is the recommended adsorptive, although argon at 77 K is also used. For carbons and graphites with low surface areas (less than about 5 m^2 g^{-1}) krypton at 77 K may be preferred as the adsorptive because of its low saturation vapour pressure. Adsorption may be carried out gravimetrically using a recording, vacuum microbalance, although a volumetric technique is generally preferred. A method for volumetrically determining the BET surface area of carbons and graphites using nitrogen at 77 K as adsorptive is given in ASTM (C819)-77.

4.1 Experimental methods

The essential features of a volumetric apparatus for surface area determination are in Figure 6.

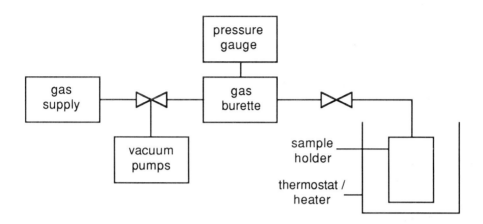

Figure 6. Schematic diagram of a volumetric adsorption apparatus.

Before adsorption of nitrogen can be measured it is necessary first to remove physically adsorbed gases from the surface of the sample by outgassing, i.e., heating under vacuum. The precise conditions for outgassing will depend upon the porous material; typical conditions for carbons are 250 °C overnight, when a residual pressure of 0.1 Pa can be achieved. To start measurements of adsorption, the outgassing heater is replaced by a liquid nitrogen thermostat; it is important to maintain a constant level of liquid nitrogen and to measure its temperature using a vapour pressure thermometer. When a steady-state condition has been achieved, nitrogen gas is admitted to the gas burette and the quantity calculated from pressure, volume and temperature (PVT) measurements using the equation of state for the gas. In modern systems the pressure gauge is usually a high-precision, wide-range, capacitance-type pressure transducer. When the gas is admitted to the sample holder the pressure falls due to expansion and to adsorption by the sample. The decrease due to expansion can be calculated from a prior dead volume calibration of the sample holder. When equilibrium is re-established, as indicated by a constant pressure reading, the amount of unadsorbed gas remaining can be calculated from another PVT measurement and hence the amount of gas adsorbed can be deduced. The molar amount adsorbed, n, and the equilibrium pressure, p, are the first data pair for an adsorption isotherm. The process is then repeated to provide further data pairs until the required pressure range has been covered. A desorption isotherm may be obtained by reversing this procedure and progressively reducing the adsorbate gas pressure by removing gas using the vacuum pumps. For porous solids adsorption and desorption isotherms are usually not reversible so that hysteresis effects are obtained; these will be discussed in the main section on mesoporous carbons. Automated instruments for measuring gas adsorption are available; most have dedicated microcomputers for experimental control, data acquisition and data analysis. In the gravimetric method the amount adsorbed at a given pressure is measured directly using a recording microbalance from which the sample is suspended; this method is also capable of automation and computer control.

A variant on the classical method for gas adsorption is the dynamic or quasi-equilibrium method. The apparatus is similar to that in Figure 6, but the adsorptive gas flows continuously into the adsorption apparatus through a mass flow controller/meter at a flow rate which is independent of the pressure in the apparatus. The pressure rise is monitored continuously using a pressure transducer and the data from the mass flowmeter and the pressure transducer are collected by a microcomputer. The pressure rise in the absence of the adsorbent can be calculated using the dead volume of the apparatus obtained by prior calibration. The actual pressure rise is less than this value due to adsorption and so the amount adsorbed, n, can be calculated. Using this method many thousands of data points can be obtained in contrast to the few tens of points obtained by the classical method. Attainment of equilibrium can be demonstrated by showing that the adsorption isotherm is independent of the flow rate of gas into the apparatus. Satisfactory results can be obtained relatively easily for non-porous, macroporous and mesoporous adsorbents. Extra care is needed when working with microporous adsorbents because long times are needed to obtain equilibrium. The dynamic method has also been adapted for use with recording microbalances to measure the amount adsorbed gravimetrically.

4.2 The Brunauer-Emmett-Teller (BET) theory

The BET theory was the first to describe successfully multilayer adsorption of gases on a wide range of porous and non-porous solids. It is for this reason that the BET equation is now accepted as the standard equation for analysing adsorption isotherms to obtain specific surface areas of solids. It should be noted that the equation has serious limitations when applied to microporous solids such as activated carbons (see the sub-section on the application of the BET equation to microporous carbons). Detailed accounts of the theory and its advantages and limitations can be found in standard textbooks (Gregg and Sing, 1982). The BET theory is based upon assumptions that the first adsorbed layer (the monolayer) is localised on surface sites of uniform adsorption energy and that second and subsequent adsorbed layers (the multilayers) build-up by a process analogous to condensation of the liquid adsorbate. In the form of the equation which is most frequently used, it is assumed that an infinite number of layers of adsorbate can be formed. A convenient, linear form of the BET equation is:-

$$\frac{p}{n(p^0 - p)} = \frac{1}{n_m c} + \frac{(c-1)}{n_m c} \frac{p}{p^0} \qquad (3)$$

where (for unit weight of adsorbate) n is the molar amount adsorbed at pressure p relative to the saturated vapour pressure, p^0, n_m, is the amount of adsorptive required to cover the surface as a monolayer (the monolayer capacity), and c is a dimensionless constant which is related to adsorption energy. To a first approximation c is given by:-

$$c = \exp\left[\frac{(q_1 - q_2)}{RT}\right] \qquad (4)$$

where $(q_1 - q_2)$ is the net heat of adsorption, q_1 being the heat of adsorption in the first adsorbed layer and q_2 the heat of adsorption in the second and subsequent layers (the molar heat of condensation). The assumption of an infinite number of adsorbed layers is reasonable for adsorption on a non-porous solid, but it is not so for adsorption on a porous solid. A form of the BET equation in which a finite number of adsorbed layers is assumed is available (Gregg and Sing, 1982), but is rarely used.

An adsorption isotherm for N_2 at 77 K for a non-porous graphitised carbon black, Vulcan 3G, which is used as a standard reference adsorbent, and the same data plotted according to Equation (3) are in Figure 7. The data conform to Equation (3) for $p/p^0 < 0.5$, but deviate from the BET equation at higher pressure.

Values of n_m and c can be obtained from data in the range of the isotherm which is linear in BET co-ordinates from the slope, s, and intercept, i, of the estimated line since $n_m = 1 / (s + i)$ and $c = s / (1 + i)$. The surface area of the carbon black can be obtained from n_m if a value for the effective cross-section area, a_m, of the adsorbate molecule can be calculated or assumed. The specific surface area, A_s, is then given by:-

$$A_s = n_m L a_m \qquad (5)$$

where L is Avogadro's number. The choice of a value for a_m has been and still is the subject of debate. An excellent summary of the arguments for selecting a value of a_m for a range of adsorbates is given by Gregg and Sing (1982). For N_2 the recommended value of a_m is 0.162 nm^2, although it appears that the value can vary with the nature of the adsorbent by up to 20 %. The recommended value of a_m yields a value of A_s = 69 m^2 g^{-1} for Vulcan 3G, in good agreement with the standard value; the value of c is 1270.

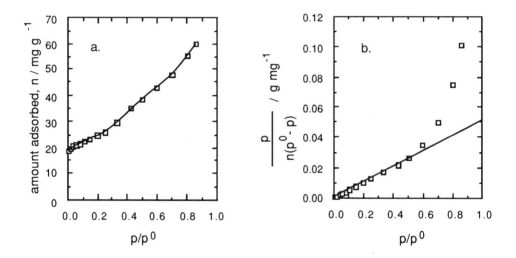

Figure 7 Adsorption of N_2 at 77 K on Vulcan 3G graphitised carbon black.
a - adsorption isotherm; b - BET plot.

4.3 Fractal surfaces of carbons

The surface of a solid may be characterised by a surface fractal dimension, D, which is a measure of surface irregularity or roughness. For a smooth surface the value of D approaches 2, the dimenson of a Euclidean surface. For a very irregular or rough surface the value of D approaches 3, the dimensions of a Euclidean volume; thus 2 < D < 3. The value of D for a surface fractal object may be estimated by measuring its area with different size yardsticks, so that increasingly small yardsticks gain access to increasingly small surface irregularities. In the case of adsorbent surfaces, the yardsticks are usually adsorptive molecules of different size, the value of D being found from the increase in A_s with the decrease in a_m. For a series of adsorbate molecules of different size, but with geometrically similar cross sections, e.g., spherical molecules, the relationship is:-

$$A_s \propto a_m^{[(2-D)/2]} \tag{6}$$

Thus D may be found from a log A_s vs log a_m plot. Avnir et al. (1983) showed that for a graphitised carbon black D is close to 2, as expected for a smooth non-porous surface. For a series of activated charcoals they showed that D decreased from about 3 to about 2 with increasing activation, suggesting that activation smoothens pore surfaces. The surface fractal dimension of a porous solid can also be estimated by measuring A_s as a function of adsorbent particle size. Using this method Fairbridge et al. (1986) obtained a value of D = 2.48 for a Syncrude coke from N_2 and CO_2 surface areas. The concept of fractal dimension provides an interesting new perspective on the properties of porous solids which is being actively developed at present.

5   SURFACE AREAS FROM SMALL ANGLE X-RAY SCATTERING

The surface area of a porous solid may also be determined using small angle X-ray scattering, SAXS. Recent developments in X-ray equipment have made this technique more widely available than before and its use is likely to increase in the future. A schematic diagram of the experimental arrangement for SAXS is in Figure 8.

Point- or slit-collimated X-rays from source 'A' impinge on the sample at 'B'. The intensity of the scattered X-rays is measured as a function of the scattering angle θ by the detector 'C'. Since X-rays penetrate the entire sample volume, SAXS gives a measure of the surface area of all pores, compared with that of open pores measured using gas adsorption.

SAXS is caused by electron density fluctuations in the scattering medium on a scale from about 0.5 - 200 nm; it is therefore applicable to micro-, meso-, and macropores. For porous solids the greatest fluctuation in electron density is likely to be at the solid-void interface, i.e., at the pore wall surface. In such a case a two-phase approximation may be made which ascribes the entire SAXS to the electron density transition at the pore wall. Electron density fluctuations within the solid material are assumed to be negligible and the electron density transition at the pore wall is assumed to be sharp and the pore wall to be smooth. It will be shown later that these assumptions are not always valid. Detailed accounts of SAXS theory are published elsewhere (Glatter and Kratky, 1982) so that only the principal, practical equations for calculating surface areas are presented here.

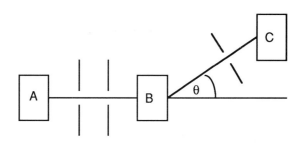

Figure 8. Schematic diagram of a SAXS apparatus.

5.1 The Debye equation

A useful method for obtaining the surface area of a porous solid was developed by Debye, Anderson and Brumberger (1957), who obtained a solution for the integral equation relating the variation in intensity, I, of X-rays scattered from a material with wave vector, $h = 4\pi \sin(\theta/2) / \lambda$, where θ is the angle between the incident and scattered X-rays and λ is the wavelength. For point-collimated incident X-rays the Debye equation may be written as:-

$$I(h) = \frac{8\pi I_e(h) a^3 \rho_e^2 V_S V_T V}{(1 + h^2 a^2)^2} \qquad (7)$$

where I_e is the intensity of X rays scattered by a single electron, ρ_e is the electron density of the solid, and V is the sample volume. The parameter a is the correlation distance, which is related to the correlation function, $\gamma(r_x)$. The correlation function is obtained by considering the probability that the ends of a randomly oriented straight line of length r_x in the material will lie in the same phase (i.e., solid or pore). The Debye equation, Equation (7), follows if the correlation function is of an exponential form, i.e.,

$$\gamma(r_x) = \exp(-r_x / a) \tag{8}$$

which is appropriate if the scattering entities (pores) are random in size, shape and location. The correlation distance, a, can be found from a plot of $I(h)^{-1/2}$ vs h^2, whose slope and intercept are s and i respectively, and $a = (s/i)^{1/2}$. The SAXS specific surface area, A_x, is given by:-

$$A_x = 4 V_S V_T / a \rho \tag{9}$$

where ρ is the density of the solid material and pores contributing to X-ray scattering.

The Debye equation for slit-collimated SAXS is given by:-

$$I(h) = \frac{8\pi^2 I_e(h) \, a^2 \rho_e^2 V_S V_T V}{(1 + h^2 a^2)^{3/2}} \tag{10}$$

In this case a is obtained from a plot of $I(h)^{-3/2}$ vs h^2, with A_x determined from Equation (9) as for point-collimation.

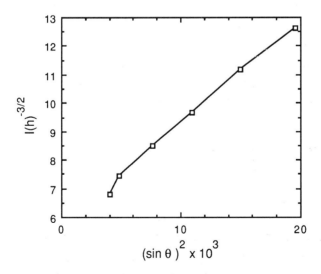

Figure 9. A Debye plot for slit-collimated SAXS from a cellulose carbon. HTT = 1270 K; I(h) in arbitrary units.

A typical Debye plot for a cellulose carbon prepared at a heat treatment temperature, HTT, of 1270 K is in Figure 9. This gives a value of a = 0.20 nm and A_x = 1 200 m² g⁻¹.

5.2 The Porod law

At large values of h, i. e., at the tail of the SAXS curve, the variation of I(h) with h approximates to Porod's law (Porod, 1952), which for point-collimation is written as:-

$$I(h) = \frac{2\pi I_e(h) \rho_e^2 S}{h^4} \qquad (11)$$

where S is the irradiated pore-wall surface area. A normalisation procedure leads to a formula for the specific surface area:-

$$A_x = \lim_{h \to \infty} [h^4 I(h)] \frac{\pi V_S V_T}{\rho \int_0^\infty h^2 I(h) \, dh} \qquad (12)$$

For slit-collimation Equation (11) becomes:-

$$I(h) = \frac{2\pi I_e(h) \rho_e^2 S}{h^3} \qquad (13)$$

and equation (12) becomes:-

$$A_x = \lim_{h \to \infty} [h^3 I(h)] \frac{\pi V_S V_T}{\rho \int_0^\infty h^2 I(h) \, dh} \qquad (14)$$

If the two-phase approximation is valid and the scattering entities are random in size, shape and location, i.e., the exponential correlation function is also appropriate, then Equations. (9) and (12) or (14) should yield similar values of A_x. This was found to be the case by Johnson and Tyson (1970) for a poly(acrylonitrile)-based carbon fibre, HTT = 2923 K. Values of A_x obtained from

Equations (9) and (12) were 190 and 200 m^2 g^{-1} respectively. These values are slightly higher than that obtained from nitrogen adsorption because of the presence of some closed pores which, by definition, are inaccessible to the adsorptive.

The two-phase approximation is likely to be acceptable for well-ordered carbons, such as the carbon fibre mentioned above and graphitised materials. For disordered carbons and those heated to low HTT, deviations from Porod's law are sometimes found. These deviations have been attributed by Schiller and Mering (1967) and by Perret and Ruland (1968) to electron density fluctuations in the <00l> direction in the solid carbon and to a gradual transition in electron density at the pore wall. For slit-collimation Ruland proposed that the deviations could be accounted for by scattering which varies as h^{-1}:-

$$I(h) = Ah^{-3} + Bh^{-1} \tag{15}$$

where A and B are constants.

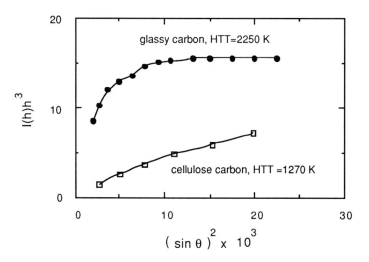

Figure 10. Examples of Porod plots for carbons.

Examples of SAXS data for two carbons as Porod plots, $I(h)h^3$ vs h^2, are in Figure 10. The plot for the glassy carbon, HTT = 2250 K, becomes horizontal at large h, thus showing classical Porod behaviour. The plot for the cellulose carbon, HTT = 1270 K, follows Equation (15); similar plots were found by Janosi and Stoeckli (1979) for activated carbons. Such plots allow the SAXS data to be corrected for the h^{-1} contribution so that A_x values can then be calculated as before.

It has been suggested that deviations from Porod's law also occur if the pore wall surface is fractal rather than smooth. Bale and Schmidt (1984) have shown that for a fractal pore wall surface:-

$$I(h) \propto h^{-(6-D)} \tag{16}$$

For a smooth surface $D = 2$ and Equation (15) applies. More generally, for a fractal surface $2 < D < 3$ and so $I(h) \propto h^{-x}$, where $3 < x < 4$. Schmidt (1988) gives a useful summary of the application of fractal theory to SAXS.

5.3 The Guinier equation

For a dilute, porous system consisting of isolated, non-intersecting pores of uniform size and shape, Guinier showed that the SAXS intensity distribution in the limit as $h \to 0$ was a Gaussian-type function, i.e.,

$$\lim_{h \to 0} I(h) = I_e(h) \rho_e^2 N_p V_p^2 \exp\left(-\frac{h^2 R_g^2}{3}\right) \tag{17}$$

where N_p is the number of pores of volume V_p, and R_g is the radius of gyration about an electronic centre of mass, or the Guinier radius. A plot of $\log_e I(h)$ vs h^2 yields R_g from the slope. The assumption of a dilute porous system is unlikely to be valid for highly porous materials, e.g., activated carbons, and Janosi and Stoeckli (1979) found for two industrial activated carbons that there was no Guiniêr limit as the wave vector approached zero. An alternative explanation is that there is a distribution of sizes of the scattering entities, and analyses are available to deal with this situation (Glatter and Kratky, 1982). To relate R_g to pore dimensions it is necessary to assume a pore shape; for example, the radius r of a spherical pore is given by $r = (5/3)^{1/2} R_g$.

6 MICROPOROUS CARBONS

Microporous carbons have a very disordered structure as revealed by high resolution electron microscopy, HREM, Figure 2, and various model structures have been proposed (Jenkins et al., 1972; Oberlin et al., 1980; Johnson, 1987). Although the models differ in detail, the essential feature of all of them is a twisted network of defective carbon layer planes, cross-linked by aliphatic bridging groups. The width of layer planes varies, but typically is about 5 nm. Simple functional groups (e. g. -OH, C=O, C-O-C) and heteroatoms are incorporated into the network and are bound to the periphery of the carbon layer planes. Functional groups can have an important influence on adsorption, but detailed consideration is outside the scope of this Chapter. In microporous carbons the layer planes occur singly or in small stacks of two, three or four with variable interlayer spacings in the range 0.34 - 0.8 nm. Fryer (1981) also found an average layer plane separation of 0.7 nm for a steam-activated anthracite carbon, and separations in the range 0.6 - 0.8 nm for Carbosieve. There is considerable microporosity in the form of an interconnected network of slit-shaped pores formed by the spaces between the carbon layer planes and the gaps between the stacks. Thus the widths of pores formed by interlayer spacings (typically from about 0.34 to 0.8 nm) are significantly less than 2 nm, the arbitrarily-defined upper limit for micropore widths (Sing et al., 1985). Constrictions in the microporous network are particular features of the ultrastructure which control access to much of the pore space. Constrictions may also occur due to the presence of functional groups attached to the edges of layer planes and by carbon deposits formed by the thermal cracking of volatiles.

6.1 Adsorption in microporous carbons

The complex and disorganised structure of microporous carbons, in which the dimensions of the pore space are commensurate with the dimensions of adsorbate molecules, makes interpretation of adsorption very difficult. First, gases are strongly adsorbed at low pressures in micropores, because there is enhancement of adsorption potential due to overlap of the force fields of opposite pore walls (see sub-section on calculations of adsorption potentials). Second, constrictions in the microporous network cause activated diffusion effects at low adsorption temperatures when the adsorptive has insufficient kinetic energy to penetrate fully the micropore space; this poses problems when using N_2 at 77 K as the adsorbate. Third, microporous carbons can exhibit molecular sieve action, i.e., the selective adsorption of small molecules in narrow micropores. Carbons also exhibit molecular shape selectivity by preferential adsorption of flat molecules, as expected from the slit-shape of micropores. Molecular sieve action can be exploited to effect separation of gas mixtures by microporous carbons, such as oxygen and nitrogen separation from air using the pressure swing method (Juntgen et al., 1981). A measure of the micropore size distribution can be obtained from the capacity of the carbon for molecular probes of different size, determined by adsorption, heats of wetting, or apparent density measurements (Spencer, 1967).

6.2 Calculations of adsorption potentials

There have been several theoretical studies of adsorption potentials in model micropores (Gurfein et al., 1970; Stoeckli, 1974; Everett and Powl, 1976). Using the Lennard-Jones 12:6 intermolecular potential function, adsorption potentials were calculated by Everett and Powl (1976) for slit-shaped pores formed between semi-infinite slabs and between single layer planes, and for cylindrical pores in an infinite slab and for those whose walls comprise a single layer of solid atoms. Pores bounded by an infinite solid give rise to 9:3 potentials, while those bounded by single atomic layers give 10:4 potentials. The 10:4 potential function for slit-shaped pores appears to be the more appropriate model, considering the structure of microporous carbons. Potentials have a single minimum for narrow pores and two minima in wide pores. Figure 11 shows the minimum 10:4 potential in the model pore, Φ^*, relative to that on the plane surface, Φ, plotted against the pore half-width, d, relative to the collision radius of the adsorptive molecule, r_0. For 10:4 potentials the two minima occur when $d/r_0 > 1.140$, Φ^*/Φ is a maximum of 2.00 at $d/r_0 = 1.00$, and Φ^*/Φ decreases to about 1.1 at $d/r_0 = 1.50$. At d/r_0 less than 1.00, i. e., between 'a' and 'b' in Figure 11; the enhancement of adsorption potential decreases rapidly due to the short range repulsive terms in the 10:4 potential function.

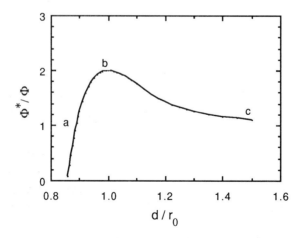

Figure 11. Normalised 10:4 adsorption potential in a model slit-shaped micropore (after Everett and Powl, 1976).

Pores with enhanced adsorption potential have been termed ultramicropores and wider pores, up to 2 nm width, have been termed supermicropores (Dubinin, 1974). The definition of the size range of ultramicropores is not precise and depends upon the size of the adsorptive molecule, but, for the purposes of illustration, may be taken as the range corresponding to 'a' - 'c' in Figure 11. For Ar (r_o = 0.340 nm) and Xe (r_o = 0.375 nm), ultramicroporous slit-widths, W_d = 2d, are 0.68 - 1.02 nm and 0.75 - 1.13 nm respectively. The width W_d is the minimum internuclear distance between C atoms in opposite walls of the model pore. If slit-widths are corrected for the finite radius of the C atoms to give W_e, the effective pore width, then for Ar and Xe, W_e = 0.24 - 0.68 nm and 0.30 - 0.79 nm respectively; these values of pore widths are in the range of interlayer spacings measured by HREM (Jenkins et al., 1972; Oberlin et al., 1980; Johnson, 1987; Fryer, 1981).

Calculated adsorption potentials in model micropores have been used to estimate critical dimensions of pores below which diffusion of gases becomes activated. For the Everett-Powl 10:4 potentials this dimension corresponds to d/r_o at point 'a' in Figure 11. The value of this dimension is critical for the ability of molecular sieve carbons to separate gases of different molecular size. Recent calculations (Rao et al., 1985; Rao and Jenkins, 1987) for diffusion through slit-shaped model micropores gave critical dimensions for several monatomic and polyatomic molecules. The value of the critical pore dimension for Ar (0.575 nm) is in very good agreement with that obtained from the Everett-Powl 10:4 potentials, and the values for N_2 and O_2 (0.572 and 0.542 nm respectively) are consistent with experiments on the separation of these gases from air (Juntgen et al., 1981).

Useful as model pore calculations are in elucidating the nature of adsorptive-adsorbent interactions in micropores, they are highly idealised when applied to microporous carbons. For example, the models ignore factors such as the highly defective nature of the carbon layer planes and edge effects resulting from their finite size, and the role of polar groups, e.g., C=O, in adsorptive-adsorbent interactions. The calculations also relate to the interaction of an isolated adsorbate molecule with the pore walls and so co-operative effects (adsorbate-adsorbate interactions) during micropore filling are not considered.

6.3 Application of the BET equation to microporous carbons

The BET equation is subject to severe limitations when applied to microporous carbons. Values of surface areas up to about 4000 m^2 g^{-1} for some highly activated carbons are unrealistically high, since the calculated surface area for an extended graphite layer plane, counting both sides, is about 2800 m² g⁻¹ Adsorption in micropores does not take place by successive build-up of molecular layers, as supposed by the BET theory. Rather, the enhanced adsorption potential in micropores induces an adsorption process described as primary or micropore filling (Dubinin, 1974). Because of its widespread use for other adsorbents, the BET surface area will continue to be used for microporous carbons, but its notional character should be recognised. The BET equation can be applied with more confidence to adsorption of gases on the non-microporous surface of carbons, i.e., on mesopores, macropores and the external surface, provided that the microporous contribution to adsorption can be effectively removed. Three types of method have been proposed: comparative methods, preadsorption techniques and isotherm subtraction.

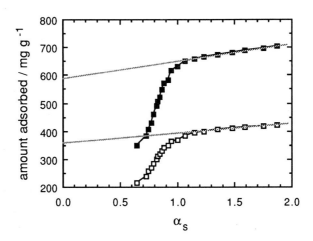

Figure 12. Typical α_s-plots for N_2 adsorption at 77 K on two activated carbons.

In comparative methods adsorption on a microporous adsorbent is compared to adsorption on a non-porous reference adsorbent of well-defined surface area. Graphitised carbon blacks have been used as reference adsorbents for carbons, but there is debate about the choice of suitable reference materials. The most widely used comparative methods are the t-plot and α_s-plot (Figure 12); these have been discussed in detail by Gregg and Sing (1982). In the former method, t is the statistical thickness of the adsorbed layer and given by $t(p/p^o) = v/A_s$, where v is the volume adsorbed at relative pressure p/p°, and A_s is the surface area ; in the monolayer, when $v = v_m$, the monolayer volume, t can be calculated by making assumptions about molecular packing. The isotherm for the non-porous reference adsorbent leads to a relationship between p/p° and t which is used to transform the conventional adsorption isotherm for a microporous adsorbent into the form v vs t. The statistical thickness t is essentially a normalising parameter for producing a reduced isotherm. Sing (1968) proposed an alternative, empirical, normalising parameter, $\alpha_s = v/(v_{p/p° = 0.4})$; t-plots and a_s-plots for microporous carbons are of similar form. Figure 12 shows α_s-plots for two activated carbons.

The steep rises at low α_s values are attributed to micropore filling and the linear regions at high α_s values to adsorption on the non-microporous surface. The micropore volume is estimated by extrapolation of the linear region to $\alpha_s = 0$ and the non microporous surface area, A_s', can be estimated from the slope of the linear portion of the α_s-plot.

In preadsorption methods micropores are filled at room temperature with a strongly adsorbed vapour, usually n-nonane, which is retained in micropores on subsequent outgassing. The non-microporous surface area is then measured by adsorption of N_2 at 77 K, while the micropore volume can be obtained by comparison of N_2 adsorption before and after adsorption of nonane. The third method of separating microporous and non-microporous adsorption is isotherm subtraction. This is a simple method for estimating A_s' from a single isotherm in which the contribution from micropore filling at p/p° > 0.1, obtained by extrapolation of the low pressure Dubinin-Radushkevich (DR) isotherm (see the sub-section on the DR equation), is subtracted from the total high pressure isotherm to give a residual, non-microporous isotherm which can be analysed by the BET equation to yield A_s'. Martin-Martinez et.al. (1986) have carried out an extensive comparison of the nonane preadsorption and isotherm subtraction methods. They conclude that for carbons with narrow micropores the two methods give very similar results; when the distribution of micropore sizes is wider the preadsorption technique fails because nonane is not retained in some wide micropores on outgassing. For superactivated carbons with a very wide range of pore sizes, neither technique was able to separate completely microporous and non-microporous adsorption.

6.4 **The Dubinin-Radushkevich (DR) equation**
Because adsorption in microporous carbons occurs by primary or micropore filling, Dubinin and his group originally modelled microporous adsorption using

the Polanyi potential theory. Dubinin found empirically that the characteristic curves for adsorption on many microporous carbons could be linearised using the Dubinin-Radushkevich (DR) equation (Dubinin, 1975), written as:-

$$V = V_0 \exp(-\varepsilon / E)^2 \qquad (18)$$

where V is the volume adsorbed at relative pressure p/p^0, V_0 is the micropore volume, E is an energy constant and $\varepsilon = RT \ln(p^0/p)$ is the adsorption potential, where R is the gas constant and T is the absolute temperature. Dubinin showed that the energy constant E could be factorised into a characteristic energy, E_0, which relates to the adsorbent, and an affinity or similarity coefficient, β, which is a constant for a given adsorptive. Essentially, β is a shifting factor which allows the characteristic curves for different adsorptives on the same adsorbent to be superimposed.

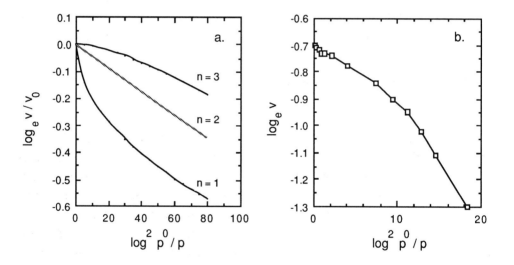

Figure 13. Deviations from the DR equation.
a - due to n ≠ 2; b - due to activated diffusion.

For some carbons the DR equation is linear over many orders of magnitude of pressure. For others, however, deviations from the DR equation are found (Marsh and Rand, 1970). In such cases the Dubinin-Astakhov (DA) equation has been proposed (Dubinin, 1975) in which the exponent two in the DR equation is replaced by a third adjustable parameter, n. Examples of the DA equation plotted in DR co-ordinates are in Figure 13a. For n > 2, the isotherm plotted in DR co-ordinates is convex to the abscissa; such deviations from the DR equation are found for well-activated carbons with a wide range of micropore sizes. For n < 2 the isotherm is concave to the abscissa and such deviations from the DR equation are found for unactivated, or slightly activated carbons with a narrow range of micropore sizes. Caution is needed in interpreting deviations from the DR equation because some are due to activated diffusion. These deviations are characterised by a downward deflection in the DR plot at low p/p^0; an example is given in Figure 13b.

Because of its empirical origins and the numerous types of deviation from it, the DR equation has not met with general acceptance. However, a justification for the DR and DA equations can be found by considering the generalised adsorption isotherm (GAI) (Ross and Olivier, 1964). The structures of microporous carbons are highly disordered and their surfaces are energetically heterogeneous so that adsorption on them may be modelled by the GAI, in which it is assumed that adsorption on sites having energy q, can be represented by a local relative isotherm $\theta(p, q)$, and where the overall relative isotherm, $\Theta(p)$, is given by

$$\Theta(p) = \int_\Omega \theta(p, q) \, f(q) \, dq \, , \ p \in \phi \tag{19}$$

where f is the site energy distribution function defined on a domain Ω and ϕ is the domain of Θ; this integral equation is the GAI. Solving the GAI for f given θ and Θ is ill-posed, that is small perturbations in Θ, e.g., experimental errors, can result in widely different solutions for f. The general approach to solving the GAI is to restrict the class of possible solutions by imposing constraints on f which correspond to assumptions about or prior knowledge of the system. The solution is then considered to be the f in this restricted class which gives the best fit to the experimentally-determined total isotherm, e.g., by least squares. One particular approach is to constrain f to be a smooth, analytic function which, with a suitable equation for θ, allows the GAI to be integrated to give an analytic function for Θ, the parameters of which can then be estimated from the data, and substituted into f to define a solution. A recent example of this approach was presented by Sircar (1984), who chose the Langmuir equation for the local isotherm and a gamma-type function for f. A solution to the GAI can also be obtained by a numerical method called regularisation (McEnaney et al., 1987; McEnaney and Mays, 1988). A third method used to solve the GAI is the condensation approximation

(CA) (Cerofolini, 1974) in which the local isotherm is approximated by a step function. The CA is a variational method which gives a solution which is exact in the limit of zero absolute temperature. Applying the CA to the Langmuir local isotherm results in the relationship $\varepsilon = q - q_0$, where q_0 is a constant, minimum adsorption energy. If f is assumed to be a Weibull distribution:-

$$f(q) = \frac{n(q - q_0)^n}{E_0^n} \exp\left\{-\left[\frac{(q - q_0)}{E_0}\right]^n\right\} \tag{20}$$

then integration of the GAI gives the DA equation, or the DR equation if $n = 2$. Thus, q_0 and the DA parameters n and E_0 represent, in an approximate way, the influence of adsorbent heterogeneity on the adsorption process. The ability of the DA and DR equations to reflect adsorption heterogeneity may explain their success in describing adsorption on a wide variety of porous and non-porous solids.

An alternative approach to the DR equation is to assume that E_0 is a characteristic energy for a homogeneous micropore structure and that heterogeneity is accounted for by assuming a distribution function for E_0, i. e., assuming that the DR equation is the local isotherm in the GAI. This approach has led to generalised DR equations such as that proposed by Dubinin and Stoeckli (1977), where $\Theta = V/V_0$ and ε is a function of E_0.

6.5 Estimation of dimensions of micropores

A useful objective for adsorption studies is the estimation of micropore sizes, e. g., for characterising and comparing different carbons. In the case of carbons exhibiting molecular sieve action, the most obvious, if laborious, method to obtain a micropore size distribution is to measure adsorption isotherms for gases of different molecular size; a recent example of this method was presented by Carrott et al. (1988). Alternatively, a single parameter estimate of micropore sizes can be obtained from the characteristic energy, E_0, of the DR and DA equations. From a comparison of E_0 with the average Guinier gyration radius, R_g, Dubinin and Stoeckli (1980) have proposed a direct inverse relationship:-

$$E_0 R_g = 14.8 \pm 0.6 \text{ nm kJ mol}^{-1} \tag{21}$$

A similar relationship was proposed between E_0 and the width of micropores accessible to molecular probes, W_m:-

$$W_m E_0 = K \text{ nm kJ mol}^{-1} \quad (22)$$

where K/E_0 is a weak function of E_0. McEnaney (1988) has shown recently that SAXS and molecular probe data may be correlated equally well with E_0 using a two-constant, semi-logarithmic equation:-

$$W_m = 4.691 \exp(-0.0666 \, E_0) \text{ nm} \quad (23)$$

Considering the limited nature of the experimental data, these estimates of micropore size must be regarded as approximate, as is emphasised by their authors.

7 MESOPOROUS CARBONS

Mesopores are in the size range, 2 - 50 nm, where adsorption and desorption are influenced by the curvature of an adsorbate meniscus formed within the pore; indeed the size range of mesopores is effectively defined by this criterion. In carbons, mesopores can be formed by enlargement of micropores, *e.g.*, by reaction with oxidising gases as in activated carbons. Also, some of the macroporous pores and cracks discussed in the Introduction may also extend into the mesoporous size range; for example, there is electron microscopical evidence for mesoporous Mrozowski cracks.

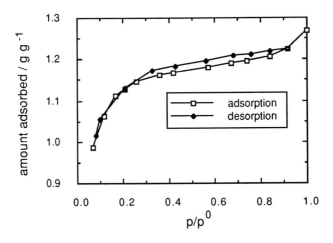

Figure 14. Adsorption-desorption isotherm of CCl_4 at 25 °C on an activated carbon.

Adsorption isotherms for mesoporous solids frequently show closed-loop hysteresis in the range p/p^0 = 0.40 - 1.00. Figure 14 is an example for adsorption of CCl_4 on an activated carbon. Hysteresis occurs because the mechanism of adsorption in mesopores is different from that for desorption. The shape of a hysteresis loop may be related to the dominant pore shape in the mesoporous solid. Hysteresis loop shapes have been classified by IUPAC (Sing et al., 1985); the type shown in Figure 14 is that found for slit-shaped pores, as expected for mesopores in activated carbons.

7.1 The Kelvin equation

Considering mesopores as capillaries, hysteresis may be understood as a capillary condensation phenomenon in which the capillary empties during desorption at a lower p/p^0 than that at which it fills during adsorption. Capillary condensation is described by the Kelvin equation which is derived from the Young-Laplace equation, and which relates the vapour pressure, p, over a concave mensicus to its principal radii of curvature, r_1 and r_2:-

$$\frac{1}{r_1} + \frac{1}{r_2} = - \frac{[RT \log_e(p/p^0)]}{\sigma V_m} \qquad (24)$$

where σ is the surface tension of the liquid and V_m its molar volume. Simple models for the capillary pores are usually assumed, e.g., non-intersecting cylinders or slits. For a meniscus formed in a cylinder $r_1 = r_2 = r_p$, the radius of the cylindrical pore; for a slit, $r_1 = d_p$ and $r_2 = \infty$ where d_p is the half-width of the slit-shaped pore. More generally $1/r_1 + 1/r_2 = 2/r_k$ where r_k is the mean or Kelvin radius, so that Equation (24) becomes:-

$$r_k = \frac{2 \sigma V_m}{RT \log_e(p^0/p)} \qquad (25)$$

Examples of pore models which lead to hysteresis are in Figure 15. During adsorption the open-ended cylindrical pore, 'a', fills at a p/p^0 value given by:-

$$p/p^0 = \exp - (\frac{2 \sigma V_m}{r_1 RT}) \qquad (26)$$

since $r_2 = \infty$. However, desorption occurs at a lower p/p^0, since it is controlled by the curvature of the hemispherical meniscus at the open ends of the pore, given by:-

$$p/p^0 = \exp - (\frac{4 \sigma V_m}{r_1 RT}) \quad (27)$$

In the case of the ink-bottle pore, 'b', Figure 15, adsorption occurs at a relative pressure controlled by r_b, the radius of the bottom of the bottle, but desorption occurs at a lower pressure controlled by r_n, the radius of the neck of the bottle. However, while both the cylindrical and ink-bottle geometries are convenient idealisations of real pore shapes, a more realistic model for mesoporous carbons is 'c' in Figure 15 in which hysteresis is influenced by random constrictions in the porous network.

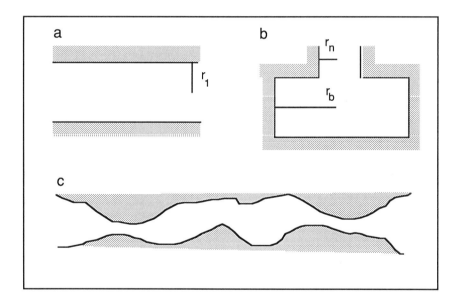

Figure 15. Examples of models for mesopores which give rise to adsorption hysteresis.

Capillary condensation in porous adsorbents does not occur on the pore wall but on an adsorbate layer whose thickness varies with p/p^0. The Kelvin radius can be corrected for the adsorbed layer using the multilayer thickness calculated from a t-plot on a suitable non-porous reference adsorbent. For a cylindrical pore the correction is obtained from $r_p = r_k + t$, and for a slit-shaped pore, $d_p = r_k + 2t$, where r_p and d_p are the radius and half width of the respective pores.

The Kelvin Equation may be used to construct a mesopore size distribution in the form $\Delta V/\Delta r$ vs Δr, where ΔV is the change in pore volume in the range Δr. Numerous methods are available for computing pore size distributions, most of which take account of the varying thickness of the adsorbed layer; the methods have been reviewed critically and in detail by Gregg and Sing (1982). Since most of the methods for calculating the pore size distribution assume a simple pore shape, it is possible to transform the pore size distribution, $\Delta V/\Delta r$ vs Δr into a pore wall surface area distribution, ΔA vs Δr, and so obtain a cumulative specific surface area, A_s.

Kiselev (1945) proposed an alternative method for obtaining pore wall surface areas from hysteresis loops. In this method the surface area destroyed by filling the pore during the adsorption branch, or created by emptying the pore during the desorption branch is computed by an integration procedure based on the isotherm plotted as $\log_e(p^0/p)$ vs n, using the equation:-

$$A_s = \frac{RT}{\sigma} \int_{n_a}^{n_b} \log_e(p^0/p)\, dn \qquad (28)$$

where the limits n_a and n_b are the amounts adsorbed at the start and end of the hysteresis loops respectively. The integration yields a higher value of A_s from the adsorption branch than the desorption branch and Kiselev proposed empirically that the mean of these two values should be used.

7.2 Limitations of the Kelvin equation
The methods for obtaining pore size distributions and surface areas using the Kelvin equation are highly developed and widely accepted. For example, some of the commercial, automatic adsorption instruments are provided with software based on one of these methods to provide pore size and pore area distributions. However, these methods are subject to some serious limitations, which have been reviewed by Gregg and Sing (1982) and more recently by Everett (1988). There are four main limitations as follows. First, the lower limit of r for applying the

Kelvin equation is a subject of debate. It is clear that the values of σ and V_m in mesopores are likely to deviate from the bulk values as pores become smaller, and ultimately the concept of a meniscus, upon which the Kelvin equation is based, will cease to be valid as pore widths approach molecular dimensions. Second, the possibility of menisci having radii of curvature of opposite sign, e.g., the meniscus which forms at the point of contact of two spheres is not considered in standard methods. Third, the t-plot method underestimates the thickness of the adsorbed surface layer, since the adsorbed layer in a pore is thicker than that on a plane surface at the same p/p^0. The fourth and most serious limitation of the Kelvin equation is the neglect in non-intersecting capillary models of pore network effects. In porous networks the behaviour of one pore can be influenced by the state of neighbouring pores. In particular, pore blocking effects can occur when a pore which is capable of emptying by capillary evaporation at a given ambient p/p^0 is prevented from doing so by neighbouring, smaller, filled pores which block access to the external surface of the carbon. The ink-bottle pore in Figure 15 provides a simple illustration of this principle, since, on desorption, the smaller pore (the neck) prevents emptying of the larger pore (the body). The result of network effects such as pore blocking is that the shape of the hysteresis loop is influenced not only by the sizes of pores, but also by their location in the solid and how they are connected.

8 MACROPOROUS CARBONS

Examples of macropores in carbons were discussed in the Introduction, and illustrated in Figures 2-5. The majority of macropores are formed during the fabrication of the carbon and graphite artefact, although they can be modified by subsequent treatments of the material. The commonest method used to characterise macropores in carbons is mercury porosimetry.

8.1 Mercury porosimetry

In capillary condensation pores fill with liquid spontaneously at a pressure less than p^0 and a meniscus which is concave to the vapour phase is formed. A concave meniscus occurs because the condensed adsorbate wets the walls of the capillary; in the Kelvin equation, Equation (24), the contact angle is implicitly assumed to be zero. The converse of this is that a liquid, such as mercury, which does not wet pore walls (i. e., contact angle > 90°) forms a meniscus which is convex to the vapour phase and can only be forced into pores by applying hydrostatic pressures greater than p^0. The pressure required to force mercury into an (assumed) cylindrical pore is related to its radius, r, by the Washburn equation (Washburn, 1921) written as:-

$$r = -\frac{2\sigma\cos\theta}{\Delta p} \qquad (29)$$

where σ is the surface tension of mercury, θ is the contact angle and $\Delta p = p^l - p^g$ is the difference between the pressure applied to the liquid mercury and that of the gas in the pore; the Washburn equation is a special case of the Young-Laplace equation (Gregg and Sing, 1982).

In porosimetry the volume of mercury penetrating the porous solid is measured as a function of applied pressure and related to the pore radius using Washburn's equation; the solid is evacuated prior to measurement so that p^g is neglected since $p^g \ll p^l$. The pore volume penetrated for a given pressure, p^l_i, is for all pores with radii $\geq r_i$, where r_i is calculated from Equation (29). Thus the results from a porosimeter are presented as a cumulative pore volume distribution with decreasing pore radius. A method for obtaining pore size distributions for porous carbons and graphites using mercury porosimetry is in ASTM (C699)-45.

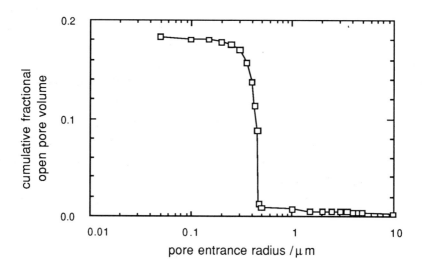

Figure 16. An example of a mercury porosimetry plot for a nuclear graphite.

Modern porosimeters operate from sub-atmospheric pressures to about 5000 atmospheres, equivalent to a pore radius range of about 7.5 μm to about 1.5 nm.

Figure 16 is an example of a pore size distribution for a nuclear graphite which was obtained by mercury porosimetry. This curve shows that the mean pore radius is about 0.4 µm and that V_o, the fractional open pore volume, is about 0.18. The very steep change in penetrated volume at about r = 0.4 µm and the flat regions either side of this radius indicate that the material has a narrow distribution of pore sizes, a characteristic of many fine-grained nuclear graphites.

Although mercury porosimetry is one of the most widely-used techniques for characterising the porous texture of solids, particularly macroporous solids, there is a number of serious limitations as detailed by Gregg and Sing (1982) and Scholten (1967). For instance, the assumed model of non-intersecting cylindrical pores is subject to the limitations of the Kelvin equation discussed earlier. Another fundamental problem concerns the values assumed for the factors of the Washburn equation: a value of contact angle $\theta = 140°$ is usually assumed, but measurements suggest that it can lie anywhere between $130°$ and $150°$. A value of surface tension $\sigma = 480$ mNm^{-1} is also assumed, but mercury is particularly prone to contamination and even small amounts of impurities can significantly change this value. For mercury, as for other liquids, the value of the contact angle may depend not only upon the physical and chemical state of the solid surface, but also upon whether the liquid is advancing or retreating over it, so that different pore volume distributions will be obtained depending on whether mercury is penetrating or withdrawing from the material. Hysteresis for porosimetry curves obtained by increasing and then decreasing the applied pressure will also be affected by network effects such as pore blocking, as discussed before for adsorption hysteresis in mesopores. For porous networks of the aperture-cavity type, which are commonly found in carbons and graphites, e.g., 'b' and 'c' in Figure 15, the critical dimension for filling a pore is that of the constricted entrance, rather than the larger dimension of the main body of the pore. It is for this reason that the radius obtained by mercury porosimetry is referred to as a pore entrance radius.

Another possibility is the modification of the porous structure of the solid at high applied pressures. Existing open pores may be prised apart and pore walls may be breached, allowing access to closed pores. Conversely, compaction of the pore structure may occur so that some pores are narrowed or even closed. In graphites it has been shown that mercury and helium pore volumes are in agreement provided that the mercury pressure is less than about 200 atmospheres, suggesting a lower limit for mercury porosimetry of about 76 nm, i.e., confined to the macropore size range. In contrast, some authors have shown good agreement between pore size distributions obtained by mercury porosimetry and capillary condensation theory, which suggest that the method can be used into the mesopore size range. It is generally accepted that the method is not suitable for characterising micropores.

8.2 Fluid transport in pores

Fluid flow through carbons depends largely upon the open pore structure. An established method for investigating open porosity is to relate measurements of fluid flow to parameters of a model pore structure (Hewitt, 1965; Grove, 1967). The most widely used pore model (once again) comprises bundles of non-intersecting cylindrical capillaries, each of constant radius and of length very much greater than the radius. This model usually results in satisfactory fits to experimental data, especially for steady-state flow. Model pore structure parameters for carbons obtained from fluid flow measurements have been used to predict flow rates for fluids under conditions different from the experiment and have been incorporated into models of structure-related processes such as oxidation (Johnson, 1981).

Two fluid transport methods are commonly used for carbons: permeation and diffusion. Permeation is the isothermal, steady-state flow of a pure fluid and a factor K, the permeability, is defined by the Carman equation (Carman, 1956):-

$$K = \frac{J\,RT\,L}{A\,\Delta p} = \frac{B_0 <p>}{\eta} + \frac{4\,K_D <v>}{3} \tag{30}$$

where J is the molar flow rate through a sample of length L and cross-section area A, Δp is the pressure difference across L, $<p>$ is the mean pressure across L, η is the fluid viscosity and $<v>$ is the mean speed of the fluid molecules. The permeability is an index of the ease with which fluids flow through a porous solid, and may be considered analogous to the thermal conductivity, i.e., an index of how easily heat is conducted. The viscous flow coefficient, B_0, which characterises the laminar (Poiseuille) flow through the capillaries is related to the transport pore volume, V_t, and the pore radius, r, by the relation:-

$$B_0 = \frac{V_t\,r^2}{24\,q^2} \tag{31}$$

while the Knudsen or slip flow coefficient, K_D, which characterises flow due to collisions between fluid molecules and the pore walls (Knudsen flow) is related to these parameters by:-

$$K_D = \frac{V_t r}{6 q^2} \qquad (32)$$

The tortuosity, $q = l/L$ is the true length, l, of the capillaries divided by their apparent length, L, through a sample. The Carman equation is generally applied to gas flow but, for liquids $K_D \to 0$ and a form of Darcy's law, written as:-

$$K = \frac{V_t r^2 <p>}{24 \eta q^2} \qquad (33)$$

has been used for determining pore structure using mercury as the liquid in a method called mercury permeametry (Hewitt, 1967). Figure 17 is a permeability plot for N_2 at 296 K for the same nuclear graphite for which porosimetry data were shown in Figure 16.

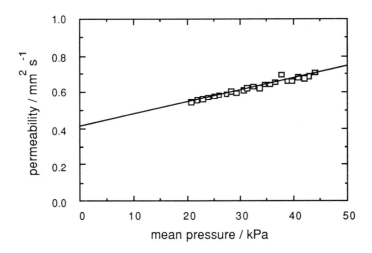

Figure 17. Permeability of N_2 at 296 K in a nuclear graphite.

Over the range of the data, it is seen that permeability increases linearly with mean pressure, as expected from the Carman equation. Using Equations (31) and (32), the pore radius is estimated as 0.4 μm, i.e., the same as that obtained from mercury porosimetry, so that it appears that gas transport is controlled by the dimensions of constrictions in the pore network, as in mercury porosimetry. The transport pore volume, V_t, is estimated to be 0.02, i.e., about 10 % of the open pore volume, V_O, obtained from mercury porosimetry.

In diffusion methods, steady-state counter-current flow rates of components of an isothermal, isobaric binary gas mixture are related to pore structure by:-

$$D_{e_i} = \frac{J_i \, RT \, L}{A \, \Delta p_i} = D_{12} \frac{V_t}{3 q^2} \quad \text{(molecular diffusion)} \tag{34a}$$

$$= \frac{2 V_t \, r \, <v>}{9 q^2} \quad \text{(Knudsen diffusion)} \tag{34b}$$

where D_{e_i} is the effective diffusivity of component i = 1 or 2 in the pores, and D_{12} is the binary molecular (free-space) diffusivity. The derivation of Equations 34 assumes that the components have similar molecular weights. Equation (34a) applies at high pressures where intermolecular collisions in the gas phase dominate flow (molecular diffusion); Equation (34b) applies at low pressures where Knudsen flow predominates. At intermediate pressures a more general diffusion equation applies of which Equations (34) are limiting forms (Mason and Malinauskas, 1983).

The relationships for permeation and diffusion can be generalised to account for pores of different sizes. Using this approach, a general diffusion equation has been developed recently to allow the estimation of pore size distributions from an analysis of the transition from molecular diffusion to Knudsen flow (Mays and McEnaney, 1985). Equation (34a) shows that molecular diffusion does not depend upon the dimensions of pores. However viscous flow [Equation (31)] and Knudsen flow [Equations (32), (34b)] increase with r^2 and r respectively, so that their coefficients are dominated by the size of macropores.

The steady-state techniques described above relate to fluid flow in transport pores, Figure 1. Methods based upon unsteady-state flow have been employed to investigate blind pores by measuring the transient elution of tracer gases from carbons and graphites (Clark et al., 1979; Clark and Robinson, 1979; Clark et al., 1983). Results are analysed using Fick's second law of diffusion to account for the time-dependecy of partial pressures.

Experimental data for fluid flow in porous carbons can generally be fitted successfully to the non-intersecting capillary models for pores described above. However, the parameters obtained from such models must be regarded as effective or notional, since the real pore structure in carbons is considerably more complicated than the model. A particular limitation of this model is its inability to predict network effects such as percolation. Pore network models have been developed to account for fluid flow, but are of limited practical use unless a great deal of very accurate flow data are available (Cunningham and Williams, 1980).

8.3 Image analysis

The limiting resolution of the optical microscope is about 1 µm, so that large macropores in carbons can be observed in flat, polished sections using reflected light, e.g., Figures 2-5. Such images contain information on two-dimensional structure parameters which can be extracted manually using point and line counting techniques, but the procedures are extremely laborious, since a large number of pores must be measured to ensure that results are statistically significant. In recent years the development of automated, quantitative image analysis has rendered manual methods obsolete (Gonzales and Wintz, 1977). The essential principles of computer-based image analysis are illustrated in Figure 18.

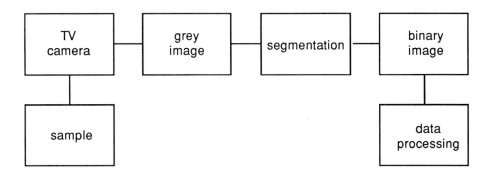

Figure 18. The principles of image analysis.

The sample to be studied is viewed by a TV camera mounted on an optical microscope. This produces an image in which contrast is revealed by a range of grey levels from black to white. The signal from the camera is directed to a computer, and the grey image is converted to a binary image by a process called segmentation. In it the features of interest are identified by a narrow range of grey levels which are selected as white in the binary image, all other grey levels being black. The binary image consists of a two-dimensional matrix of picture elements, 'pixels', which are either switched on in the white regions or off in the black regions. This image can then be processed by a computer. While the application of image analysis to porous carbons and graphites has been largely confined to optical microscopical images of macropores, in principle the method may also be applied to mesopores and micropores using electron microscopical images.

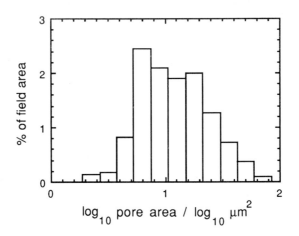

Figure 19. A pore area distribution for a glassy carbon obtained from image analysis.

An object in a binary image (a pore) is uniquely defined by a list of pixel co-ordinates along its boundary. from which various geometrical factors may be calculated, such as area, A, perimeter, P, and maximum and minimum chord lengths, l_{max}, l_{min}, between two boundary points (Feret diameters). From these parameters other size factors may be defined, e.g., the radius of the circle with the same area as the feature, $r_c = (A/\pi)^{1/2}$, a shape parameter, the aspect ratio, l_{max}/l_{min}, and orientation parameters, e.g., the angle between l_{max} and a reference axis. As an example of image analysis data, Figure 19 shows a distribution of pore area for bubble pores in a glassy carbon of the type illustrated in Figure 3.; this carbon has a narrow range of pore sizes with a mean area of about 10 μm^2. Image analysis also shows that the aspect ratio of these pores is close to 1, as expected from visual examination of the micrographs.

The main advantage of quantitative image analysis is that, unlike the methods discussed previously in this Chapter, a model pore structure does not initially have to be assumed. Also, image analysis of macropores in carbons can yield structural data which it would be difficult or impossible to obtain from other methods. Image analysis can be used to characterise the different types of open and closed pores in carbons and graphites, discussed in the Introduction, provided they can be differentiated, e.g., by different grey levels or by their shape. Pore shape factors obtained from image analysis have been used in empirical models for strength-structure relationships in metallurgical cokes (Patrick and Walker, 1986) and to demonstrate that pores of high aspect ratio are preferentially developed in thermal oxidation of graphites (Burchell et al., 1986). Another useful application of image analysis is the study of local variations in porosity within a sample. This approach has been applied to a study of the strength of glassy carbon beams (Pickup et al., 1986) where it was shown that the area fraction of pores varied with distance across the beam; this variation influenced the bend strength of the material. In a recent study (Yin et al., 1989) the orientations of cracks in needle-coke particles were used to determine the orientation of filler grains with respect to the extrusion direction in an electrothermal graphite. It was shown that orientation varied strongly with radial position in the graphite log, being greatest at the edge and least at the centre. This has implications for the radial variation of electrical conductivity in the log and hence to its performance as an electrode.

The principal limitation of the image analysis method is that the information obtained relates to a two-dimensional image of the object. Structural information in three-dimensions can be obtained from image analysis by the laborious technique of serial sectioning, i.e., by analysing a large number of adjacent sections through the feature of interest. Alternatively, stereological techniques may be applied (Underwood, 1970) in which a model is assumed which relates the two-dimensional image to a three-dimensional structure. Stereology often involves difficult mathematical problems, but for structurally simple systems, e.g., a random distribution of spherical pores, it can yield useful data such as pore volumes. A random distribution of spherical pores is a realistic model for the macropores in the glassy carbon in Figure 3, and in this case the application of a stereological theorem known as Delesse's principle to the image analysis data gave the area fraction of pores, $A_f = V_T = 0.07$. The assumption of simple pore structures is often made implicitly in image analysis, but can be misleading when dealing with anisometric or highly oriented features.

References

Avnir, D., Farin, D. and Pfeifer, P. (1983). J. Chem. Phys. **79**, 3566.

Bale, H. D. and Schmidt, P. W. (1984). Phys. Rev. Lett. **53**, 596.

Burchell, T. D., Pickup, I. M., McEnaney, B. and Cooke, R. G. (1986). Carbon **24**, 534.

Carman, P. C. (1956). Flow of Gases Through Porous Media, Butterworths London.

Carrott, P. J. M., Roberts, R. A. and Sing, K. S. W. (1988). Stud. Surf. Sci. Catal. **39**, 89.

Cerofolini, G. .F. (1974). Thin Solid Films **23**, 129.

Clark, J. D. and Robinson, P. J. (1979). J. Mat. Sci. **17**, 2649.

Clark, J. D., Ghanthan, C. S and Robinson, P. J. (1979). J. Mat. Sci. **14**, 2937.

Clark, J. D., McVey, M. D., Murad, S. and Robinson, P. J. (1983). J. Mat. Sci. **18**, 593.

Cunningham, R. E. and Williams, R. J. J. (1980). Diffusion in Gases and Porous Media, Plenum Press, New York.

Debye, P., Anderson, H. R. and Brumberger, H. (1957). J. Appl. Phys. **28**, 679.

Dubinin, M. M. (1974). J. Colloid Interface Sci. **46**, 351.

Dubinin, M. M. (1975). Prog. Surf. Membr. Sci. **9**, 1.

Dubinin, M.M. and Stoeckli, H.F. (1977). J. Colloid Interface Sci. **59**, 184.

Dubinin, M. M. and Stoeckli, H. F. (1980). J Colloid Interface Sci. **75**, 34.

Everett, D H (1988). Stud. Surf. Sci. Catal. **39**, 1.

Everett, D. H. and Powl, J. C. (1976). J. Chem. Soc., Faraday Trans. **172**, 619.

Fairbridge, C., Ng, S. H. and Palmer, A. D. (1986). Fuel **65**, 1759.

Fryer, J. R. (1981). Carbon **18**, 347.

Glatter, O. and Kratky, O. (Eds.) (1982). Small Angle X-ray Scattering, Academic Press, London.

Gregg, S. J. and Sing K. S. W. (1982). Adsorption, Surface Area and Porosity, 2nd ed., Academic Press, London.

Grove, D. M. (1967). In: Porous Carbon Solids, R. L. Bond, Ed., pp 203-224, Academic Press, London.

Gonzales, R. C. and Wintz, P. (1977). Digital Image Processing, Addison-Wesley, London.

Gurfein, N. S., Dobychin, D. P. and Koplienko, L. S. (1970). Russ. J. Phys. Chem. **4**, 411.

Hewitt, G. F. (1965). Chem. Phys. Carbon **1**, 73.

Hewitt, G. F. (1967). In: Porous Carbon Solids, R. L. Bond, Ed., pp 203-224, Academic Press, London.

Janosi, A. and Stoeckli, H. F. (1979). Carbon **17**, 465.

Jenkins, G. M., Kawamura, K. and Ban L. L. (1972). Proc. Roy. Soc. **A327**, 501.

Johnson, D. J. (1987). J. Phys. D: Appl. Phys. **20**, 286.

Johnson, D. J. and Tyson, C. N. (1970). J. Phys. D: Appl. Phys. **3**, 526.

Johnson, P. A. V. (1981). Nuclear Energy **20**, 231.

Juntgen, H., Knoblauch, K. and Harder, K. (1981). Fuel **60**, 2472.

Kiselev, A. V. (1945). Usp. Khim. **14**, 367.

Marsh, H. and Rand, B. (1970). J. Colloid Interface Sci. **33**, 101.

Martin-Martinez, J-M, Rodriguez-Reinoso, F., Molina-Sabio, M., and McEnaney, B. (1986). Carbon **24**, 255.

Mason, E. A. and Malinauskas, A. P. (1983). Gas Transport in Porous Media: The Dusty Gas Model, Elsevier, Amsterdam.

Mays, T. J. and McEnaney, B. (1985). Proc. 17th Biennial Conference on Carbon, Lexington, KY, USA, pp 104-105.

McEnaney, B. (1988). Carbon **26**, 267.

McEnaney, B. and Mays, T. J. (1988). Stud. Surf. Sci. Catal. **39**, 151.

McEnaney, B., Mays, T. J. and Causton, P. D. (1987). Langmuir **3**, 695.

Mrozowski, S. (1956). Proc. 1st and 2nd Conferences on Carbon, Buffalo, NY, USA, pp. 31-45, Waverly Press, Boston.

Oberlin, A., Villey, M. and Combaz, A. (1980). Carbon **18**, 347.

Patrick, J. W. and Walker, A. (1986). Proc 4th International Carbon Conference, Baden-Baden, FRG, pp 156-158.

Perret, R. and Ruland, W. (1968). J Appl. Crystallogr. **1**, 308.

Pickup, I. M., Mays, T. J. and McEnaney, B. (1986). Proc 4th International Carbon Conference, Baden-Baden, FRG, pp 240-242.

Pickup, I. M., McEnaney, B. and Cooke, R. G. (1986). Carbon **24**, 535.

Porod, G. (1952) Kolloid Z. **125**, 51.

Rao, M. B. and Jenkins, R. G. (1987). Carbon **25**, 445.

Rao, M. B., Jenkins, R. G. and Steele, W. A. (1985). Langmuir **1**, 137.

Ross, S. and Olivier, J. P. (1964). On Physical Adsorption, Interscience, New York.

Schiller, C. and Mering, C. R. (1967). C. R. Acad. Sci. Paris **B264**, 247.

Schmidt, P. W. (1988). Stud. Surf.Sci. Catal. **39**, 35.

Scholten, J. J. F. (1967) In: Porous Carbon Solids, R. L. Bond, Ed., pp. 225-249, Academic Press, London.

Sing, K. S. W. (1968). Chem. Ind. (London) February, 1528.

Sing, K. S. W., Everett, D. H., Haul, R. A. W., Moscou, L., Pierotti, R. A., Rouquerol, J., and Siemieniewska, T (1985). Pure Appl. Chem. **57**, 603.

Sircar, S. (1984). J Colloid Interface Sci. **98**, 306.

Spencer, D. H. T. (1967). In: Porous Carbon Solids, R L Bond, Ed., pp 87-154, Academic Press, London.

Stoeckli, H. F. (1974). Helv. Chim. Acta **57**, 2195.

Sutherland, J. W. (1967). In: Porous Carbon Solids, R. L. Bond, Ed., pp. 1-63, Academic Press, London.

Underwood, E., E., (1970). Quantitative Stereology, Addison-Wesley, Boston.

Walker, P. L., Jr., Rusinko, F. and Austin, L. G. (1959). Adv. Catal. **11**, 133.

Washburn, E. W. (1921). Phys. Rev. **17**, 273.

Yin, Y., McEnaney, B. and Mays, T. J. (1988). Carbon **27**, 113.

Chapter 6

Carbon Fibres: Manufacture, Properties, Structure and Applications

D.J. Johnson

Dept. of Textile Industries, University of Leeds, Leeds, LS2 9JT, U.K.

Summary.

The preparation of carbon fibres from polyacrylonitrile (PAN) is described in some detail and carbon fibres from mesophase pitch (MP) are introduced. The general properties of carbon fibres and carbon-fibre reinforced plastics (CFRP), together with some of their aerospace and other end-uses, are also included but the emphasis in this Chapter is on the structure-property relationships of PAN-based and MP-based carbon fibres. The theoretical tensile modulus and strength is compared to those values currently available commercially and attention is drawn to the remarkable improvements in tensile strength which have followed research-led understanding and improved processing.

Methods for structural investigation considered in detail are: wide-angle X-ray diffraction, especially the measurement of size parameters such as L_c, $L_{a\perp}$, and $L_{a//}$, and the orientation parameters q and Z; small-angle X-ray diffraction and the various methods for estimating pore-size parameters; scanning electron microscopy (SEM), and transmission electron microscopy (TEM) utilizing bright-field, dark-field and lattice-fringe imaging modes. Typical TEM results are illustrated and it is shown how these have led to useful models of structure and, hence, to mechanisms of fracture failure.

SEM evidence from PAN- and MP-based carbon fibres failing in tension, in flexure and in approximately uniaxial compression show how the sheet-like nature of MP-based fibres, as opposed to the highly folded and interlinked nature of the PAN-based fibres, promotes crack propagation and thus inhibits the development of good tensile and compressive strengths.

CARBON FIBRES: MANUFACTURE, PROPERTIES, STRUCTURE AND APPLICATIONS

D.J. Johnson

Department of Textile Industries, University of Leeds, Leeds, LS2 9JT, U.K.

1. INTRODUCTION

1.1 History

A purely chronological account of the history of carbon fibres will cite Edison's use of carbonised rayon for the filaments of electric-light bulbs as the first commercial use of the material, but the true history of the carbon fibres we are concerned with today begins in the 1950s and 60s with the requirement of the aerospace industry for better lightweight materials, and the realisation that low density fibres of high modulus could be used as the reinforcing elements in composites. In response to this demand, there were a number of relatively successful attempts to prepare carbon fibres, especially those of Roger Bacon at Union Carbide using viscose rayon (see Bacon 1975), of Shindo (1961) in Japan using polyacrylonitrile (PAN), and of Otani (1965), also in Japan, using an isotropic pitch. However, these procedures incorporated an expensive hot stretching process, and although fibres were produced commercially from viscose rayon (Union Carbide) and petroleum pitch (Kureha Kagaku), they have now been superseded.

Any historical account is inevitably biased by the perspective of the author; for me, the development of a successful commercial process for carbon fibres is firmly associated with the late William Watt and his colleagues at the Royal Aircraft Establishment (RAE) in Farnborough, England. It is said that in 1963 Leslie Phillips referred Bill Watt to 'black Orlon' which could be prepared from the commercial PAN fibre by heating at 200°C for several hours in air, then pyrolysing in a flame. Consequently, together with his long term colleague the late William (Bill) Johnson, Bill Watt began the first experiments with the Courtaulds PAN fibre Courtelle to bring the exothermic oxidation reaction under control. Eventually, fibres were produced which when carbonized at 1000°C had a Young's modulus of 150 GPa, and when further heat treated at 2500°C developed an even greater stiffness of 380 GPa. An understanding of the need to restrict the PAN fibres during oxidation was one of the most important developments which led to the RAE British patent for producing high modulus carbon fibres. Later the early batch process was developed into a continuous process, and in 1966 used on fully commercial production lines at Morganite Ltd. and Courtaulds Ltd. (see Mair and Mansfield 1987, Watt et al, 1966).

At the same time, scientists and engineers at Rolls-Royce Ltd. were also developing a batch process for carbon fibre production, and were experimenting with carbon fibre reinforced plastic (CFRP) as the material for the turbine blades of the then new RB 211 engine. It is well known that the bird-strike problem of delamination due to a low interlaminar shear strength eventually ruled out the use of CFRP in turbine blades, and for some time this was felt to be a major setback to the carbon fibre industry. Fortunately, the demands of sports enthusiasts are not as rigorous, they wish to bolster their imperfect skills with the best equipment available, and CFRP was the answer, thus the industry was maintained for some years by non-aerospace end uses.

With hindsight it can be seen that it was important that confidence in the use of CFRP for aerospace applications should have been brought about by very careful development and test programmes. An account by Anderson (1987) of British Aerospace PLC describes the history of their major technology programmes for CFRP starting in 1969 with the flight of a rudder trim tab for a Jet Provost and terminating in 1983 with flight testing of Tornado tailerons and a structural test to failure of a Jaguar wing. That CFRP is now accepted for the primary structures of both military and civil aircraft is a result of the ability of the material to meet very stringent property requirements. The improved quality of the reinforcing carbon fibres has been effected by synergy between process technology and research in structure/property relationships.

1.2 General properties

We have established that carbon fibres were developed primarily as a low density high stiffness (high Young's modulus) reinforcing material. Later we will discuss in detail the problem of the relatively low tensile strength and low strain-to-failure in this essentially brittle material. The compressive strength of carbon fibres is also of great importance and can vary considerably depending on fibre type; indeed the fatigue and ageing of CFRP materials may well be determined more by its compressive properties than its tensile properties; this is a major factor to be considered by designers.

The high modulus of all carbon fibres is due to good orientation of the turbostratic graphite layer planes which constitute the material. These well aligned layers of carbon atoms also give rise to good thermal and electrical conductivity, which are very useful properties for different applications. The stability of CFRP structures is assisted by a very low coefficient of thermal expansion and further enhanced by excellent damping characteristics, chemical inertness, and biocompatibility.

2. PREPARATION

Carbon fibres have been prepared from a multitude of precursor materials ranging from natural materials such as wool and lignin, to the high performance organic fibre Kevlar, but only two are of any commercial significance, PAN and mesophase pitch.

2.1 Carbon fibres from PAN

The most important conditions of the RAE process for producing carbon fibres from a PAN fibre precursor were (i) oxidation at a temperature which does not allow catastrophic runaway of the exothermic reaction, (ii) restriction of length shrinkage or even stretching of the fibres during oxidation, (iii) oxidation evenly throughout the section of the fibre.

In practise, oxidation is usually carried out for several hours at around 220°C. The pendant nitrile groups of PAN become crosslinked to form a ladder polymer, initiation of this process being promoted by the presence in Courtelle of a small amount of itaconic acid as copolymer. Oxygen is incorporated into the ladder polymer according to a number of possible schemes as described by Watt and Johnson (1975). The stabilized structure, which contains about 11wt% oxygen can then be heated in an inert atmosphere such as nitrogen without melting. In a continuous temperature gradient situation many complex reactions take place with the evolution of volatile products. In the range up to 450°C HCN, acrylonitrile, propionitrile, NH_3 and H_2O, are believed to come from reactions involving the unladdered sections of the chains; around 500°C there is another HCN peak and around 700°C another H_2O peak, believed to come from reactions involving cross-linking of the chains.

Evolution of nitrogen starts at about 700°C so that fibres produced at 1000°C have around 5.8wt% nitrogen and about 50 wt% of the mass of the original PAN precursor fibre. Continuing pyrolysis up to 1500°C eliminates most of the residual nitrogen and completes the conversion of the PAN molecules into sheets of carbon with a graphitic structure of bonds. Details of the conversion process are summarised in the excellent biography of Bill Watt by Mair and Mansfield (1987).

Considerable improvement in the alignment of the graphitic layer planes and their stacking size, with a concomitant increase in Young's modulus, can be achieved by heat treatment in the region of 2500°C. Fibres produced at these high temperatures were originally referred to as Type I, now the letters HM (High Modulus) are in vogue. Fibres treated around 1000°C were referred to as Type II , now more usually HS (High Strength or High Strain to failure; fibres treated at intermediate temperatures were referred to as Type III or Type A, now more often IM (Intermediate Modulus or InterMediate strain to failure).

2.2 Carbon fibres from mesophase pitch

Although isotropic pitch can be made into carbon fibre by melt spinning into fibre form, oxidation to render the fibres infusible, and carbonization at 1000°C, hot stretching at temperatures in excess of 2500°C is required in order to obtain the high preferred orientation of the layer planes essential for high modulus. This difficult and costly procedure eliminates the price advantage of isotropic pitch as a precursor material.

Singer (1978) has described experiments which led to Union Carbide's mesophase-pitch-based carbon fibre process; mesophase fibres drawn from acenaphthylene pitch at 450°C could be heat treated without an oxidative thermosetting step and were anisotropic with a high degree of axial preferred orientation. A comprehensive account of the preparation of carbon fibres from both isotropic and mesophase pitches, including the chemistry of the reactions leading to the graphitic sheets of carbon, can be found in the book by Donnet and Bansal(1984).

3. TENSILE PROPERTIES

3.1 Tensile modulus

The theoretical tensile or Young's modulus of graphite is usually estimated as around 1000 GPa, a considerable advantage over organic polymers which cannot exceed around 300 GPa. It was established some time ago that a good correlation exists between preferred orientation and Young's modulus in carbon fibres; it is simply that, the more highly oriented the layer planes, the higher the tensile modulus (Fourdeux et al, 1971). In order to achieve a parallel arrangement of layer planes, considerable energy in some form must be given to carbon fibres from PAN, which are essentially non-graphitizing; less energy is required to align the layer planes in a graphitizing material such as a carbon fibre from mesophase pitch, and higher moduli can be obtained with these materials. Two procedures which can give a greatly increased tensile modulus to PAN-based carbon fibres are stress graphitization at high temperature (650 GPa) (Johnson J.W. et al, 1969) and boron-doping (550 GPa) (Allen et al, 1970). Unfortunately, these methods are not commercially viable.

3.2 Tensile strength

Unlike the Young's modulus, the theoretical tensile strength of a solid is more difficult to evaluate (Macmillan, 1972). One approach is the Orowan-Polanyi expression

$$\sigma_t = (E\zeta_a/a)^{\frac{1}{2}} \tag{1}$$

which relates the theoretical strength σ_t to Young's Modulus E, the surface energy ζ_a, and the interplanar spacing a of the planes perpendicular to the tensile axis. The ratio σ/E varies considerably, but for many materials is in the range 0.1 to 0.2. For a perfect graphite, a theoretical strength of 100 GPa might be expected, but because of defects, the practical strength is always an order of magnitude lower than the theoretical strength. Thus in the most perfect form of fibrous carbon so far produced, the graphite whisker, E is about 680 GPa and σ about 20 GPa, a σ/E ratio of 0.03.

In brittle solids the defects are small cracks which act as stress concentrators; they will grow under the action of a stress σ if they are greater than a critical size C as determined by the Griffiths relationship:

$$\sigma^2 = 2E\mathcal{E}_a/\pi C \qquad (2)$$

If this relationship is applied to a graphite whisker, using the generally accepted value for $\mathcal{E}_a$ of 4.2 Jm^{-2}, the critical flaw size is 4.5 nm; for a Type I carbon fibre (σ = 2.8 GPa, E = 370 GPa) it is 126 nm, and for a Type II or Type A fibre (σ = 5.7 GPa, E = 300 GPa) it is 25 nm.

In practise, the major limitation on tensile strength has been the presence of gross flaws. This was demonstrated by the classic work of Moreton and and Watt (1974) on clean-room spinning of the PAN precursor, and the later work on contaminated mesophase-pitch precursors by Jones, Barr and Smith (1980). There have been many statistical studies of the effect of flaw distribution on tensile strength, particularly in terms of the gauge-length dependence of this property. Extrapolations to a very short gauge length give values ranging from 7 GPa (Chwastiak et al., 1979) down to 3.2 GPa (Beetz 1982). Since fibres are in production which have realisable strengths above the lower figure, we may conclude that it is now possible to reduce the flaw content to a very low level and that other factors are responsible for the strength limitations of carbon fibres.

3.3 Practical properties of carbon fibres

The tensile properties for a representative selection of commercially produced carbon fibres are given in Table 1. It is now possible to achieve a Young's modulus of around 500 GPa in HM PAN-based carbon fibres and a modulus as high as 800 GPa in a mesophase-pitch based carbon fibre, a value approaching the theoretical modulus of 1000 GPa. Unfortunately, high modulus fibres have a low strain to failure; a high strain to failure is necessary for many purposes, particularly CFRP. There have been significant improvements with HS and IM fibre types in recent years and strains to failure of around 2% can now be obtained with a commensurate tensile strength of 5 to 7 GPa.

Table 1. Tensile properties of carbon fibres. Manufacturers' data.

Manufacturer	Fibre	E (GPa)	σ (GPa)	e (%)
PAN-based High Modulus (Low strain to failure)				
Celanese	Celion GY-70	517	1.86	0.4
Hercules	HM-S Magnamite	345	2.21	0.6
Hysol Grafil	Grafil HM	370	2.75	0.7
Toray	M 50	500	2.50	0.5
PAN-based Int. Modulus (Int. strain to failure)				
Celanese	Celion 1000	234	3.24	1.4
Hercules	IM-6	276	4.40	1.4
Hysol Grafil	Apollo IM 43-600	300	4.00	1.3
Toho Beslon	Sta-grade Besfight	240	3.73	1.6
Union Carbide	Thornel 300	230	3.10	1.3
PAN-based High Strain to failure				
Celanese	Celion ST	235	4.34	1.8
Hercules	AS-6	241	4.14	1.7
Hysol Grafil	Apollo HS 38-750	260	5.00	1.9
Toray	T 800	300	5.70	1.9
Toray	T 1000	294	7.06	2.4
Mesophase-pitch-based				
Union Carbide	Thornel P-25	140	1.40	1.0
	P-55	380	2.10	0.5
	P-75	500	2.00	0.4
	P-100	690	2.20	0.3
	P-120	820	2.20	0.2

E   Young's Modulus.
σ   Tensile Strength.
e   Strain to failure.

4 STRUCTURE

Probably the most effective techniques to elucidate structure in carbon fibres are X-ray diffraction, both wide-and small-angle, and electron microscopy, both scanning and transmission. We have now reached a situation where Type I PAN-based carbon fibres and HM mesophase-pitch-based carbon fibres have been well characterised, and where we have considerable knowledge of structural features and failure mechanisms which limit tensile strength. How far this understanding applies to Type II (HS), Type A (IM), and other mesophase-pitch-based fibres is open to question. Nevertheless, improved all round understanding has prompted a considerable increase in the strain-to-failure and tensile strength of available fibres.

Here, we will briefly summarise the essential structures of carbon fibres as discovered by the application of X-ray diffraction and electron microscopy and show how the physical properties are related to the inherent structure. This treatment is selective rather than comprehensive, but provides a foundation for an understanding of the basic science of carbon-fibre structure.

4.1 Wide-angle X-ray diffraction

Wide-(or high-)angle X-ray diffraction is the simplest and quickest technique for the structural characterisation of any material. Accounts specific to the analysis of carbon-fibre patterns have been published by Fourdeux et al (1971) and Johnson (1987). Typical wet-spun PAN fibres have two main reflections on the equator, no reflections on the meridian, and no layer-line reflections. The orientation of the molecular chains is low and there is no true unit cell, only pseudo-hexagonal packing at a spacing of about 0.51 nm. After pre-stretching, there is a great improvement in orientation and the appearance of very diffuse first layer-line scattering. It is not possible to produce three-dimensional order of the molecules in PAN fibres by stretching.

After carbonization above $1000°C$, PAN-based carbon fibres exhibit a typical X-ray diffraction pattern with a broad 002 reflection on the equator and a diffuse 100 ring, strongest on the meridian, Figure 1. The small-angle scatter can also be seen but there are no 3D reflections as found in graphite. After heat treatment at $2500°C$, Type I carbon fibres show much sharper 002 and 100 reflections, and the 004 reflection is evident on the equator, Figure 2; again there are no reflections characteristic of three-dimensional order such as 101 or 112. Essentially, PAN-based carbon fibres are non-graphitizing with a turbostratic (irregular) organisation of the layer planes. In contrast, mesophase-pitch-based carbon fibres are much more graphitizing and high-temperature treated fibres show evidence of a strong 101 reflection outside the 100 ring, Figure 3, indicating the more perfect arrangement of the layer planes.

There is also evidence for preferred a-axis orientation of the layer planes with the corner of the hexagon along the fibre axis, see Figure 4. This point has been investigated quantitatively by Ruland and Plaetschke (1985), who have also shown that PAN-based carbon fibres have a-axis orientation along the fibre axis.

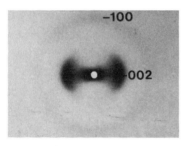

Figure 1. Wide-angle X-ray diffraction pattern of Type II (HS) PAN-based carbon fibre.

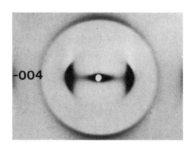

Figure 2. Wide-angle X-ray diffraction pattern of Type I (HM) PAN-based carbon fibre.

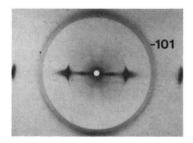

Figure 3. Wide-angle X-ray diffraction pattern of mesophase pitch-based carbon fibre (P120).

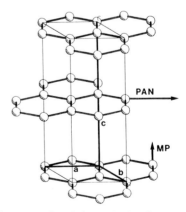

Figure 4. Unit cell of graphite showing preferred direction of the layer planes with respect to the fibre axis in PAN-based (PAN) and mesophase-pitch-based (MP) carbon fibres.

Although much information can be gained by a qualitative study of the X-ray diffraction pattern, serious characterisation necessitates more rigorous quantitative analysis. By means of an X-ray diffractometer, usually employed in step-scan mode, it is possible to obtain intensity data on the diffraction peaks in both equatorial and meridional directions. The measure of peak width B_{002} from the 002 reflection is used to measure the stacking size Lc by means of the relationship $Lc = K\lambda/B_{002}\cos\theta$, where θ is the Bragg angle and λ the X-ray wavelength. K, the Scherrer parameter, can be considered as the factor by which the apparent size must be multiplied to give the true size, and for 00ℓ reflections is generally given the value of 1. The 100 reflection is spread into a ring because of the turbostratic packing; the meridional width is used to obtain the layer-plane length parallel to the fibre axis $La_{//}$; similarly, the equatorial width is used to obtain the layer-plane dimension perpendicular to the fibre axis $La_{\perp}$. Both these measures should be considered as essentially 'correlation lengths' relating only to short perfect distances in the fibre as 'seen' by the X-rays. In fact, transmission electron microscopy shows that the layer-plane lengths and widths are very much greater than the La values measured by X-ray diffraction. K values for hk0 reflections (traditionally 1.84 for $La_{\perp}$) have been evaluated by Ruland and Tompa (1972) as a function of the preferred orientation.

An X-ray diffraction trace around a 00ℓ peak in an azimuthal direction can be used to obtain quantitative information concerning the preferred orientation of the layer planes. Two useful parameters are Z the azimuthal width at half-peak height, and q the orientation factor introduced by Ruland and Tompa (1968). This factor is somewhat complex mathematically, but in essence, a fibre with no orientation will have a q factor of 0, and a fibre with layer planes oriented perfectly along its axis will have a q factor of -1.

When the diffraction peaks overlap, it is necessary to separate them by computational methods. A full account of all the steps required to resolve the peaks and evaluate crystallite size can be found in Hindeleh et al (1983) and a short account specific to carbon fibres in Bennett et al, (1976). In brief, overlapping peaks are resolved into mixed Gaussian-Cauchy profiles which may be either symmetric or asymmetric about the peak position. The computational procedure is a minimisation of the least sum of squares between observed and calculated data in terms of parameters defining peak-profile shape, peak height, peak width, and peak position. Parameters can be constrained within set limits when necessary. Each 00ℓ peak is a convolution of size-broadening and distortion-broadening components. Numerous mathematical methods have been proposed for the separation of size and disorder using both transform and non-transform methods, see Hindeleh et al, (1983), where results from real and simulated profiles are given after evaluation by different methods. Typical values of X-ray diffraction parameters for various types of PAN-based carbon fibres are listed in Table 2. It should be noted that the lattice disorder σ_l, which is a root-mean-square value for the lattice distortion $\delta d/d$ cannot be measured for Type II or Type A fibres, because there is only one 00ℓ reflection.

Table 2. Structural parameters of PAN-based carbon fibres obtained by X-ray diffraction.

Fibre	c/2	Lc	$La_\parallel$	$La_\perp$	σ_l	Z	-q	l_p
	nm	nm	nm	nm		deg		nm
Type I	0.35	5.3	9.8	8.0	0.9	20	0.78	2.5
Type II	0.36	1.7	3.9	2.7	-	41	0.55	1.4
Type A	0.37	1.2	3.2	1.9	-	44	0.51	0.9

Lc   crystallite size in c direction (stacking size).
•$La_\parallel$ crystallite size in a direction parallel to fibre axis.
$La_\perp$ crystallite size in a direction perpendicular to fibre axis.
σ_l lattice order (cannot be measured for Type II and A fibres.
Z   preferred orientation.
q   preferred orientation factor.
l_p pore size perpendicular to fibre axis.

4.2 Small-angle X-ray diffraction

When structural units in a material scatter X-rays they do so at small angles of scatter and special instrumental techniques are necessary to obtain information about this small-(or low-)angle diffraction. If discrete reflections are observed in the small-angle pattern, then the repeat of the structure 'l_s' can be evaluated by the application of Bragg's law, although this must be considered as a first approximation which can be fully solved by more rigorous mathematics. It is best to consider the long spacing 'l_s' as a measure of a periodic density difference.

An account of the different types of small-angle pattern normally found in fibres has been included in a recent review (Johnson, 1987). Carbon fibres have a very strong equatorial lobe-shaped pattern, the development of which is illustrated in Figure 5 where small-angle patterns recorded from fibres heat-treated to different temperatures are shown. The faint discrete spot (not visible in Figure 5) and equatorial streak of the precursor PAN changes to a more intense streak with increasing temperature and is gradually replaced by the lobe-shaped carbon fibre pattern. Small-angle patterns showing the full development of the lobe-shaped scatter in Type II and Type I fibres are shown in Figure 6. The best interpretation of this pattern, and that which is in good agreement with TEM evidence, is that there is a fine structure of crystallites enclosing long needle-shaped voids and that the crystallites have a distribution of sizes (see Figure 10). The effect of improved orientation can be seen in the decreased spread of the lobe-shaped small-angle X-ray scatter of Figure 6 as heat-treatment temperature is increased.

Figure 5. Small-angle X-ray diffraction patterns from PAN fibres heat treated to different temperatures: (a) 300°C, (b) 500°C, and (c) 700°C.

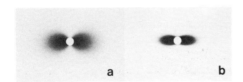

Figure 6. Small-angle X-ray diffraction patterns from PAN-based carbon fibres: (a) Type II (HS), (b) Type I (HM).

If it is necessary to obtain a measure of the void or pore size, for a powder specimen, a 'mean chord intercept length' in the pores 'l_p' can be measured using the Debye method (Johnson and Tyson, 1969,1970). A mean chord intercept length in the crystallites 'l_c' can also be evaluated. Plots of $I^{-1/2}$ against θ^2 are made and the slope-intercept ratio found to give the so-called 'correlation length' a. If the ratio of the density of the fibre to that of perfect graphite is known, then the correlation length can be used to find l_p, l_c and the specific surface S_v. The relationships used are:

$$l_p = 2a \qquad (3)$$
$$l_c = a(1 - c) \qquad (4)$$
$$S_v = 4c/l_c \qquad (5)$$

where c is the density ratio of carbon fibre to perfect graphite.
Although reasonable approximations can be made for other fibres these relationships are only completely valid for Type I fibres.

The distinction between a sharp boundary from crystallite to pore, and a boundary which might contain disordered layer planes, can be tested by Porod's law. For a powder this is that a plot of $I\theta^4$ against 2θ reaches a constant limit for a sharp boundary, and oscillates about that limit for a boundary containing disordered layers. For a fibre specimen, the law requires an $I\theta^3$ against 2θ relationship. Perret and Ruland (1968,1969) have shown that it is better to plot $I\theta^3$ against $(2\theta)^2$ for fibres. In this case, the plots tend to a straight line and do not oscillate; the part of the curve due to disordered layers can then be removed. By suitable corrections and normalisation, values of l_p can be obtained.

Values of l_p measured perpendicular to the fibre axis are included in Table 2. Other more straightforward, but somewhat approximate, methods can be used for small-angle scattering; for example, Guinier plots of log I against $(2\theta)^2$ can be used to give a radius of gyration R for the crystallite or void, and plots of log I against log θ^2 can be used to give parameters for the distribution of void size (Tomizuka and Johnson 1978).

4.3 Scanning electron microscopy

The scanning electron microscope (SEM) is the workhorse microscope in materials science; results are obtained rapidly and usually have sufficient information content for deductions of structural significance. Also, additional instrumentation is often attached to the SEM, e.g. for elemental analysis using energy dispersive X-ray analysis. Certainly, in the field of carbon fibres, the use of the SEM for structural studies is widespread. In particular, fracture-face studies by SEM are invaluable in our understanding of structure-property relationships. Nevertheless, it is important to distinguish between a structure which is the result of fracture and a structure which is inherent. Figure 7 is a micrograph of a fracture face in a Type II (HS) PAN-based carbon fibre. The radiating structure is typical of the so-called 'hackle' zone of any brittle fracture but is not typical of the inherent structure. A cut bundle of mesophase-pitch based carbon fibres is shown in Figure 8. The sheet-like structures seen in this case are most probably true structural features. Interpretation of the normal surface of a carbon fibre might appear relatively simple; it would be tempting to interpret the highly folded surface of the Type II carbon fibre in Figure 7 as evidence of a fibrillar structure; we shall see later that this is not the case.

4.4 Transmission electron microscopy

The transmission electron microscope (TEM) is another very important tool for investigating the structure of carbon fibres. Early work concentrated on thin fragments of material obtained by grinding or ultrasonic dispersion; more recent investigations have used longitudinal and transverse sections of fibres prepared by the difficult technique of ultramicrotomy. Operation of the TEM can utilise one of three imaging modes, bright-field, dark-field, or lattice-fringe; the electron-diffraction mode is also useful.

The bright-field mode is the normal mode of operation, but the dark-field mode is being used increasingly in all carbon work since it allows positive identification of the regions in a specimen which contribute to a particular reflection in the diffraction pattern. In dark-field mode, one of the diffraction spots is centred and all other reflections excluded by means of an aperture, so that the image is formed only from those crystallites diffracting electrons into the beam selected. Figure 9 is a typical dark-field (002) micrograph of a longitudinal section from a PAN-based Type I carbon fibre. The diffracting crystallites are seen as white regions on a dark background.

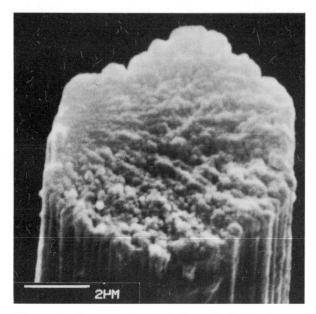

Figure 7. Fracture face of Type II (HS) PAN-based carbon fibre after tensile failure.

Figure 8. Cut bundle of mesophase-pitch-based carbon fibres.

In a TEM with a low spherical aberration coefficient (e.g. C_s = 1.6 mm), it is possible to obtain lattice-fringe images of the 0.34 nm turbostratic graphite layers at high magnification. If the imaging is carried out with the undiffracted (zero-order) reflection centred, the image is an 'axial' or 'multiple-beam' image; if the zero-order beam and a 002 diffracted beam are tilted at equal angles about the electron optical axis of the microscope, and all other diffracted beams excluded, then we have a 'tilted-beam' image. Figure 10 illustrates multiple beam lattice-fringe images from HM (a) and IM (b) PAN-based carbon fibres.

The electron diffraction mode is particularly useful when the longitudinal or transverse structure has zones of different structural organisation. Using a 'selected area' aperture it is possible to select any region and record its selected-area diffraction pattern. This pattern can be digitised and the diffraction peaks analysed by analogous methods to those used in high-angle X-ray diffraction. The characterisation of the skin-core characteristics of a typical PAN-based carbon fibre using TEM methods is described by Bennett and Johnson (1979).

A complete account of high-resolution lattice-fringe imaging in TEM, complete with computer simulations, has been provided by Millward and Jefferson (1976). An earlier account referring to PAN-based carbon fibres was given by Johnson and Crawford (1973). It has to be emphasised that there is no exact one-to-one correspondence between layer planes in the specimen and lattice fringes in the image, as can be observed both by changing the lattice image from 'multiple-beam' mode to 'tilted-beam' mode, and by changing focus, (see Figures 12 and 14 of Johnson 1987). The lattice-fringe image, indeed any image, is formed by the transfer of information from the object to the image via the objective lens, a process which is affected by the spherical aberration coefficient and the level of focus of the objective lens. With a C_s of 1.6 mm or greater, the 0.34 nm graphite layer plane spacing comes in a zone of the phase contrast transfer function (PCTF) where small changes in focus cause considerable change in the contrast of the lattice fringes. The best results in high-resolution TEM are obtained when the PCTF is as flat as possible (Scherzer focus); in other words, the maximum amount of information is transferred with the same contrast.

Figure 9. Dark-field (002) image of longitudinal section of HM PAN-based carbon fibre.

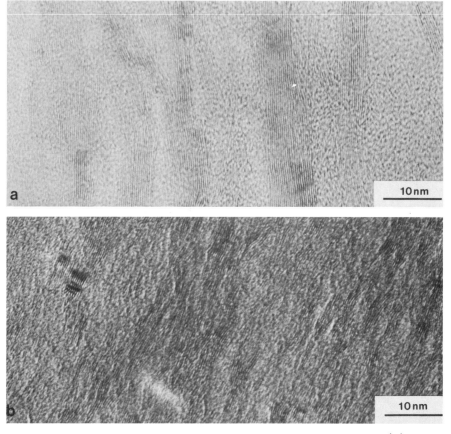

Figure 10. Lattice-fringe images of longitudinal sections of (a) HM PAN-based carbon fibre, (b) IM PAN-based carbon fibre.

A simple test for the PCTF can be made with an optical diffractometer (or with an image analyser having Fourier transform capability); any information gaps in image transfer are revealed as troughs between the peaks in the optical diffraction pattern (i.e. the Fourier transform). If, for example, gaps occur for spacings of the order of 0.5 to 0.7 nm, then there is no contrast in the image for such repeats. This can give a disordered carbon the appearance of a structure containing voids, and is particularly important when interpreting the images of Type II and Type A PAN-based carbon fibres. Crawford and Marsh (1977) have demonstrated this effect and show how valid and invalid images can be distinguished.

4.5 Microstructure

Early TEM observations on fragments of carbon fibre were held to indicate the presence of a fibrillar structure; subsequent studies of high-resolution lattice-fringe images have shown this not to be the case (Bennett and Johnson 1979). A 'ribbon-like' model in which curvilinear layer planes are packed side by side enclosing voids of an approximately needle shape, has been proposed for PAN-based carbon fibres (Fourdeux et al, 1971), Figure 11, and has some merit for relating Young's modulus and preferred orientation. A more realistic model of the structure of a Type I PAN-based carbon fibre is given in Figure 12, being a simplified representation of the complex three-dimensional interlinking of layer planes forming crystallites which enclose sharp-edged voids as seen in typical lattice-fringe images such as Figure 10 (a).

Lattice-fringe images of transverse sections reveal a very complicated situation. In the skin region, the layer planes are essentially parallel to the surface, but in addition to the type of complex crystallite interlinking seen in longitudinal section, many layer planes fold through angles of up to 180° in a 'hairpin' fashion. In the lateral direction throughout the core the layer planes are folded extensively, providing coherence over large cross-sectional areas of the fibre. Primarily the structural organisation in cross section is random; a convenient model of structure for Type I fibres is shown in Figure 13. If layer planes with different degrees of disorder are substituted in Figures 12 and 13, then we can envisage models of structure for all types of PAN-based carbon fibres.

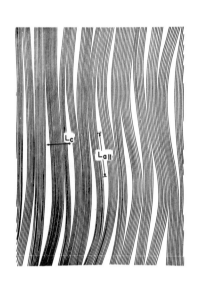

Figure 11. Ribbon-like model of structure for Type I PAN-based carbon fibres.

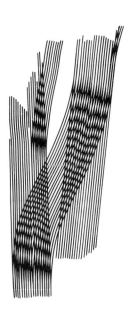

Figure 12. Schematic two-dimensional representation of longitudinal structure.

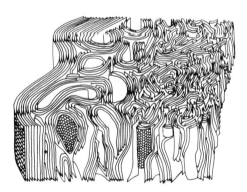

Figure 13. Schematic three-dimensional structure in a Type I PAN-based carbon fibre.

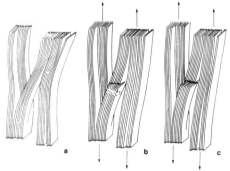

Figure 14. Reynolds-Sharp mechanism of tensile failure.
(a) Misoriented layers linking two crystallites parallel to fibre axis.
(b) Tensile stress applied parallel to fibre axis causes layer-plane rupture in direction $La_\perp$, crack develops along La and Lc.
(c) Further application of stress causes complete failure of misoriented crystallite. Catastrophic failure occurs if the crack size is greater than the critical size in either the Lc or $La_\perp$ direction.

After detailed dark-field and high-resolution TEM studies of several Type I PAN-based carbon fibres, Oberlin and her colleagues (Guigon et al, 1984) produced an alternative model in which the layer planes have a variable transverse radius of curvature which decreases continuously from the surface to the centre. They show that, as the average radius of curvature of the layer planes increases, so the tensile strength decreases. The tensile strength of other types of PAN-based carbon fibre is said to depend on a so-called 'compactness index'.

Their model is essentially one of folded and crumpled sheets of layer planes, which although entangled, are not considered to be interlinked in any way. Our model is dismissed as improbable on the grounds of the unfolding of lamellae which occurs after various treatments. However, it is unlikely that the relatively small number of bonds interlinking adjacent layer stacks, as amply demonstrated in several lattice-fringe images, will hinder exfoliation, and the model is valid. Without interlinking, the behaviour of carbon fibres in compression would probably be even worse with a mechanism of failure similar to that found with Kevlar (Dobb et al, 1981).

5   FRACTURE MECHANICS
5.1   Tensile failure

Difficulties in describing fracture mechanisms in terms of dislocation pile-up at grain boundaries, the unbending of curved ribbons, the presence of density fluctuations, or yield processes involving local shear deformation and slippage, were discussed in a review by Reynolds (1973). Although a wide range of internal defects had been observed, no simple relationship could be found between flaw diameter, fibre strength, and surface free energy. This led to the proposal by Reynolds and Sharp (1974), of a crystallite shear limit for fibre fracture.

This mechanism of fracture is based on the idea that crystallites are weakest in shear on the basal planes. When tensile stress is applied to misoriented crystallites locked into the fibre structure, the shear stress cannot be relieved by cracking or yielding between basal planes. The shear strain energy may be sufficient to produce basal-plane rupture in the misoriented crystallite, and hence a crack which will propagate both across the basal plane and, by transference of shear stress, through adjacent layer planes. A schematic diagram of a misoriented crystallite well locked into the surrounding crystallites, is shown in Figure 14(a). When stress is applied, basal plane rupture takes place, Figure 14(b), and proceeds throughout the local region, Figure 14(c). However, before a crack can propagate through a fibre and cause failure, either one of two conditions must be fulfilled.

(1) The crystallite size in one of the directions of propagation of a crack, that is either L_c or $L_{a\perp}$, must be greater than the critical flaw size C for failure in tension (C may be around 120 nm for a Type I fibre).

(2) The crystallite which initiates catastrophic failure must be sufficiently continuous with its neighbouring crystallites for the crack to propagate.

The first condition is not normally fulfilled because both L_c and $L_{a\perp}$ are much less than C, although the effective value of $L_{a\perp}$ is considerably greater than the values measured by X-ray diffraction since layer planes are curved or 'hairpin' shaped. The second condition is most likely to be satisfied in regions of enhanced crystallization and misorientation observed around defects.

In a fairly recent study of tensile failure (Bennett et al, 1983), specimens from an old batch of Type I PAN-based carbon fibres containing many flaws, were stressed to failure in glycerol. This enabled the fracture ends to be preserved intact for subsequent examination, first by SEM, and then, after embedding and sectioning, by TEM. Internal flaws which did not initiate failure were seen to have walls containing crystallites arranged mainly parallel to the fibre axis. Internal and surface flaws which did initiate failure often showed evidence of large misoriented crystallites in the walls of the flaws. Continuity of crystallites then gave rise to values of $L_{a\perp}$ which exceeded the critical flaw size.

Further proof for the concept that large misoriented crystallites, together with continuity of structure in the walls of flaws, cause fibre failure under stress, is found in an earlier study of lignin-based carbon fibres (Johnson et al, 1975). These fibres, which had very inferior tensile strengths, contained many flaws in the form of inclusions caused by catalytic graphitization around impurity particles in the precursor material (Figure 15 is a typical example). These inclusions are usually

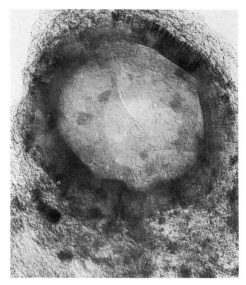 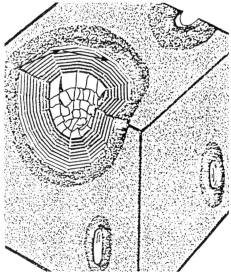

Figure 15. TEM image of longitudinal section of lignin-based carbon fibre showing graphitic inclusion

Figure 16. Schematic representation of graphitic inclusion of 15. Arrow represents propagation of crack

contiguous with the normal structure; under stress, misoriented crystallites will fail by the Reynolds-Sharp mechanism, the crack will propagate around the wall of the inclusion, and total failure will result. This is explained in the schematic diagram of Figure 16.

5.2 Flexural failure

Although considerable attention has rightly been paid to the tensile deformation of carbon fibres, much less attention has been focussed on deformations which involve the compression or combined compression and tension of a fibre. DaSilva and Johnson (1984) have caused a number of commercially available carbon fibres to be stressed to failure under both tensile and flexural deformation (knot test).

The fracture faces of the circular PAN-based carbon fibres after flexural failure were distinctly different from the faces after tensile failure. Flexural fracture faces from both HM (Type I) and HS (Type II) fibres exhibit a rough striated area and a relatively smooth but corrugated area

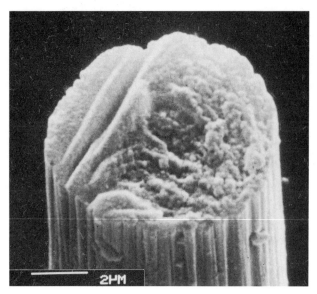

Figure 17. SEM image of fracture face of HM-Grafil fibre after flexural failure.

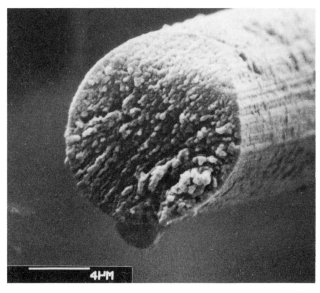

Figure 18. SEM image of fracture face of Thornel mesophase-pitch-based carbon fibre after flexural failure.

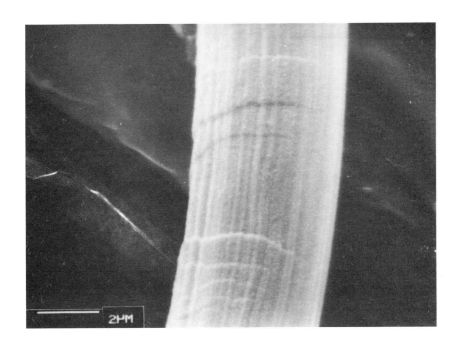

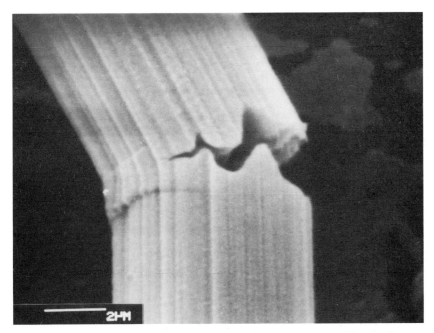

Figure 19. SEM images along typical PAN-based carbon fibre after compressive failure: (a) Kink bands seen near to the anchor zone, (b) secondary fracture face forming in anchor zone.

corresponding to the regions of the fibre under tension and compression respectively. The corrugated area probably reflects the shear stresses in the compression region. Figure 17 is a good example of a flexural failure fracture face in an HM fibre.

The fracture surface of a Thornel mesophase-pitch-based fibre, Figure 18, is remarkably similar to the surface produced by tensile failure, and to faces produced simply by cutting the fibre (Figure 8). Sheet-like features are seen in all cases; they will propagate a transverse crack much better than a random structure because of the large effective crystallite size in the direction of the sheet. The uniformity of structure seen in a knot fracture face may indicate the simultaneous failure of sheets through that part of the section in tension. When the fibres are deformed into a knot, it is inevitable that many more crystallites are misoriented; consequently, the chances of layer-plane rupture and subsequent fracture are increased.

It is most interesting that a recent study of mesophase-pitch-based carbon fibres spun into multilobal shape shows that they have relatively greater tensile strengths than the more uniformly circular fibres (Edie *et al*, 1986). Presumably sheet-like layer-plane development is restricted under these spinning conditions.

5.3 **Compressive strength**

Allen (1987) has pointed out the limitations of flexural tests for evaluating compressive strength, and has developed a recoil test to measure compressive strength in single fibres. The values obtained from a range of fibres, including two carbon fibres, were in good agreement with values obtained from composite tests. Park has further developed this recoil test for carbon fibres and we hope to publish results fairly soon (Dobb, Johnson and Park 1988). It is exciting that the effect of recoil can be observed in the SEM; Figure 19 shows the remains of a typical PAN-based fibre after compressive failure; this is the first example of carbon fibre buckling we have observed in this laboratory.

6 **APPLICATIONS**

Although we are primarily concerned with the science of carbon fibres, and particularly with the relationships between structure and physical properties, we must consider the end uses of carbon fibres and how the properties are exploited. As the major usage of carbon fibres is in CFRP, we will briefly examine how far the advances in physical properties of the fibres have been translated into advances in the properties of the composite.

6.1 Composite properties

Typical values for some of the physical properties of unidirectional composites comprising Torayca carbon fibres are given in Table 3. Along the direction of the fibres the tensile strengths and tensile moduli are very approximately halved, the strain-to-failure being about the same. Effectively, the matrix contributes very little to the longitudinal properties of the composite.

Compressive strength is difficult to measure in the fibres alone, but can be measured relatively easily in composites. It is surprising that the compressive strength of the composite utilising the very high strain-to-failure T1000 fibre is very similar to that of the lower strain-to-failure T300 fibre. We must assume that, whereas the compressive strength of T300 fibre is about 90% of the tensile strength, the compressive strength of T1000 fibre is only 50% of the tensile strength. The latter value is typical of mesophase pitch-based materials and may be compared with the poor compressional properties of Kevlar (see Dobb and Johnson 1987).

An interesting review of the matrix materials used for composites, and the properties of unidirectional and multidirectional laminates is given by Dorey (1987), who concisely summarises current research on the fibre-matrix interface. Bond strengths between fibre and matrix are optimised by a combination of physical and chemical effects dependent on the structure of the fibre surface, the method and degree of surface treatment, and the resin properties at or near the interface.

6.2 Aerospace uses

We have referred earlier to the steady progress of CFRP from stiffness critical secondary aircraft structures, such as control surfaces, to strength critical structures, such as tail fins, tail planes and wing structures. In military aircraft this includes the Tornado taileron, the Jaguar wing, and Harrier tailplanes and rudders. Now the European Fighter Aircraft (EFA), which first flew in experimental form in August 1986, will

Table 3. Comparison of properties of fibres and composites (Comp.), Manufacturer's data (GPa).

Property	T300		T1000		M30		M50	
	Fibre	Comp.	Fibre	Comp.	Fibre	Comp.	Fibre	Comp.
Tensile Strength	3.5	1.8	7.1	3.5	3.9	2.2	2.5	1.1
Compressive Strength	-	1.5	-	1.6	-	1.4	-	0.7
Tensile Modulus	230	137	294	165	294	158	490	268
Strain to failure (%)	1.5	1.3	2.4	2.0	1.3	1.3	0.5	0.4

contain 20% of airframe mass as CFRP, about 1000 kg. A similar amount of CFRP is now used in the Boeing 767 for secondary structures. The Airbus A310 and A320 are the first civil aircraft to use CFRP in primary structures (tail fins and wings), together with extensive use in secondary structures.

CFRP is also used extensively in small civil aircraft. The Lear Fan 2100 is well known, using CFRP in its fuselage and wings. Less well known is the Leopard, a four seater turbojet which uses carbon to stiffen glass composite; it is interesting to note the laments of the designer concerning the poor compressive properties of CFRP (Kent 1986). Stiffening of glass fibre composite rotor blades in helicopters is another application, with the use of CFRP in the primary structures of helicopters of considerable importance.

Carbon fibre reinforced carbon (CFRC) brakes were first developed for Concorde and the Mirage 2000 fighter; they are now finding increasing usage because of their light weight and ability to absorb energy at high temperatures, particularly in emergency situations where metal-based brakes would melt. The almost zero coefficient of linear thermal expansion of CFRP is of great importance for the stability of both aerospace antennae and satellite structures, indeed increasing usage is expected in this demanding area of aerospace.

6.3 Non-aerospace uses

The high stiffness at low weight of CFRP is of great importance in many applications concerned with sporting goods. Golf clubs, tennis, squash, and badminton rackets, vaulting poles, skis and ski poles, and fishing rods, are some of the items where CFRP helps to improve the performance of enthusiasts. Bicycles, both frames and wheel disks, surf boards, kyaks, rowing sculls and larger racing craft, have all benefited from the use of CFRP. In motor sport, CFRP racing car bodies provide lightness, stiffness, and in the event of accident, greatly improved protection for the driver. The CFRP monocoque chassis was introduced in 1981 and is said to maintain its rigidity much better than aluminium (Clarke 1986); here, good fatigue and creep resistance are important.

The fatigue resistance, low inertia, and high damping characteristics of CFRP find use in such diverse areas as high-speed machinery and audio equipment. The biological inertness and low weight of CFRP is a great advantage for prostheses, particularly artificial limbs. There are other uses of CFRP in medicine, particularly where the high X-ray transmission can be exploited. The chemical inertness of carbon can be fully utilised in carbon fabrics, but the matrix component of CFRP can be too vulnerable for many uses in the chemical industry.

6.4 Future trends

Intensive work on vapour deposited fibres has led to materials which have properties equivalent to commercially available carbon fibres, and with a relatively highly graphitic structure (see Tibbetts and Beetz 1987). The intercalation of elements such as potassium and bromine into these and other graphitic carbon fibres can lead to greatly improved conductivity. No doubt many interesting developments will result from areas of research and development concerned with more exotic forms of carbon, but there is still much to be done to realise the full potential of the conventional carbon fibres produced from PAN or mesophase-pitch precursors.

References.

Allen, S.R. (1987). Tensile recoil measurement of compressive strength for polymeric high performance fibres. J. Mater. Sci. 22, 853-859.

Allen, S., Cooper, G.A., Johnson D.J. and Mayer R.M. (1970). Carbon Fibres of High Modulus. Proc. Third London Int. Conf. Carbon and Graphite 1970, Soc. of Chemical Industry, London, 456-461.

Anderson, B.W. (1987). The impact of carbon fibre composites on a military aircraft establishment. J. Phys. D: Appl. Phys. 20, 311-314.

Bacon, R. (1975). Carbon Fibres from Rayon Precursors. Chemistry and Physics of Carbon (Eds. Walker P.L. and Thrower P.A.), Vol. 9, Marcel Dekker, New York, 1-101.

Beetz, C.P. Jr. (1982). A Self-Consistent Weibull Analysis of Carbon Fibre Strength Distributions. Fibre Sci. Technol. 16, 81-94.

Bennett, S.C., Johnson D.J. and Montague, P.E. (1976). Electron Diffraction and High-Resolution Studies of Carbon Fibres. Proc. Fourth London Int. Conf. Carbon and Graphite 1974, Soc. of Chemical Industry, London, 503-507.

Bennett, S.C. and Johnson, D.J. (1979). Electron-Microscope Studies of Structural Heterogeneity in PAN-Based Carbon Fibres. Carbon 17, 25-39.

Bennett, S.C., Johnson, D.J. and Johnson W. (1983). Strength-structure relationships in PAN-based carbon fibres. J. Mater. Sci. 18, 3337-3347.

Chwastiak, S., Barr, J.B. and Didchenko, R. (1979). High Strength Carbon Fibres from Mesophase Pitch. Carbon 17, 49 53.

Clarke, G.P. (1986). The Use of Composite Materials in Racing Car Design. Carbon Fibres Technology, Uses and Prospects, (Ed. The Plastics and Rubber Institute, London), Noyes, Park Ridge, NJ. 137-145.

Crawford, D. and Marsh, H. (1977). High resolution electron microscopy of carbon structure. J. Microscopy 109, 145-152.

DaSilva, J.A.G. and Johnson, D.J. (1984). Flexural studies of carbon fibres, J. Mater. Sci. 19, 3201-3210.

Dobb, M.G. and Johnson, D.J. (1987). Structural Studies of Fibres Obtained from Lyotropic Liquid Crystals and Mesophase Pitch. Developments in Oriented Polymers -2 (Ed. Ward, I.M.), Elsevier, London and New York, 115-152.

Dobb, M.G., Johnson, D.J. and Park, C.R. (1988). Compressional properties of High Performance Fibres. (In preparation).

Dobb, M.G., Johnson, D.J. and Saville, B.P. (1981). Compressional behaviour of Kevlar fibres. Polymer 22, 960-965.

Donnett, J.B. and Bansal, R.P. (1984). Carbon Fibres. Marcel Dekker, New York.

Dorey, G. (1987. Carbon fibres and their applications. J. Phys. D: Appl. Phys. 20, 245-256.

Edie, D.D., Fox, N.K., Barnett, B.C. and Fain, C.C. (1986). Melt-Spun Non-Circular Carbon Fibres. Carbon 24, 477-482.

Fourdeux, A., Perret, R. and Ruland W. (1971). General structural features of carbon fibres. Proc. First Int. Conf. Carbon Fibres, Plastics Institute, London, 57-62.

Guigon, M., Oberlin, A. and Desarmot, G. (1984). Microtexture and Structure of Some High-Modulus, PAN-Base Carbon Fibres. Fibre Sci. Technol. 20, 177-198.

Hindeleh, A.M., Johnson, D.J. and Montague, P.E. (1983). Computational Methods for Profile Resolution and Crystallite Size Evaluation in Fibrous Polymers. Fibre Diffraction Methods, ACS Symposium No. 141 (Fds. French, A.D. and Gardner, K.H.), American Chemical Society, Washington, 1983, 149-182.

Johnson, D.J. (1987). Structural Studies of PAN-Based Carbon Fibres. Chemistry and Physics of Carbon (Ed. Thrower P.A.), Vo. 20, Marcel Dekker, New York, 1-58.

Johnson, D.J. and Crawford, D. (1973). Defocussing phase contrast effects in electron microscopy. J. Microscopy 98, 313-324.

Johnson, D.J., Tomizuka, I. and Watanabe, O. (1975). The Fine Structure of Lignin-based carbon fibres. Carbon 13, 321-325.

Johnson, D.J. and Tyson C.N. (1969). The fine structure of graphitised fibres. J. Phys. D: Appl. Phys. 2, 787-795.

Johnson, D.J. and Tyson, C.N. (1970). Low-angle X-ray diffraction and physical properties of carbon fibres. J. Phys. D: Appl. Phys. 3, 526-534.

Johnson, J.W., Marjoram, J.R. and Rose, P.G. (1969). Stress Graphitization of Polyacrylonitrile Based Carbon Fibres. Nature 221, 357-358.

Jones, J.B., Barr, J.B. and Smith, R.E. (1980). Analysis of flaws in high-strength carbon fibres from mesophase pitch. J. Mater. Sci. 15, 2455-2465.

Kent, D.W. (1986). Design and Fabrication of Two Sophisticated Hand layed-up Carbon Fibre Composite Products. Carbon Fibres Technology, Uses and Prospects (Ed. The Plastics and Rubber Institute, London), Noyes, Park Ridge, NJ. 137-145.

Macmillan, N.H. (1972). The Theoretical Strength of Solids. J. Mater. Sci. 7, 239-254.

Mair, W.N. and Mansfield, E.H. (1987). William Watt 1912-1985. Biographical Memoirs of Fellows of the Royal Society 33, 643-667.

Millward, G.R. and Jefferson, D.A. (1976). Lattice Resolution of Carbons by Electron Microscopy. Chemistry and Physics of Carbon (Eds. Walker P.L. and Thrower P.A.), Vol. 14, Marcel Dekker, New York, 1-82.

Moreton, R. and Watt, W. (1974). Tensile Strengths of Carbon Fibres. Nature 247, 360-361.

Otani, S. (1965). On the Carbon Fibre from the Molten Pyrolysis Products. Carbon 3, 31-38.

Perret, R. and Ruland, W. (1968). X-ray Small Angle Scattering of Non-Graphitizable Carbons. J. Appl. Cryst. 1, 308-313.

Perret, R. and Ruland, W. (1969). Single and Multiple X-ray Small-Angle Scattering of Carbon Fibres. J. Appl. Cryst. 2, 209-218.

Reynolds, W.N. (1973). Structure and Physical Properties of Carbon Fibres Chemistry and Physics of Carbon (Eds. Walker, P.L. and Thrower P.A.) Vol. II, Marcel Dekker, New York, 1-67.

Reynolds, W.N. and Sharp, J.V. (1974). Crystal Shear Limit to Carbon Fibre Strength. Carbon 12, 103-110.

Ruland, W. and Plaetschke, R. (1985). Preferred Orientation of the Internal Structure of Carbon Layers in Carbon Fibres. Seventeenth Biennial Conf. Carbon, Lexington American Carbon Soc. and University of Kentucky 356-357.

Ruland, W. and Tompa, H. (1968). The Effect of Preferred Orientation on the Intensity Distribution of (hk) Interferences. Acta Cryst. A24, 93-99.

Ruland, W. and Tompa, H. (1972). The Influence of Preferred Orientation on the Line Width and Peak Shift of (hk) Interferences. J. Appl. Cryst. 5, 225-230.

Shindo, A. (1961). Osaka Kogyo Gijitsu Shikenjo Koho 12, 110-119.

Singer, K.S. (1978). The Mesophase and High Modulus Carbon Fibres from Pitch. Carbon 16, 409-415.

Tibbetts, G.G. and Beetz, C.P. Jr. (1987). Mechanical properties of vapour-grown carbon fibres. J. Phys. D: Appl. Phys. 20, 292-297.

Tomizuka, I. and Johnson, D.J. (1978). Microvoids in Pitch-based and Lignin-based Carbon Fibres as Observed by X-ray Small-angle Scattering. Yogyo-Kyokai-Shi 86, 42-48.

Watt, W., Phillips, L.N. and Johnson, W. (1966). High-Strength High-Modulus Carbon Fibres. The Engineer 221, 815-816.

Watt, W. and Johnson, W. (1975). Mechanism of oxidisation of polyacrylonitrile fibres. Nature 257, 210-212.

Chapter 7

Mechanical Properties of Cokes and Composites

J.W. Patrick and D.E. Clarke

Carbon Research Group, Loughborough Consultants Ltd., Dept of Chemical Engineering, University of Technology, Loughborough, Leicestershire, LE11 3TF, U.K.

Summary.

In any application where a solid material is subjected to an applied load, the response of the material to the loading is determined primarily by its mechanical properties. Cokes, carbons and carbon composites are a diverse group of materials in terms of origin, production and resultant structure but show similarities in their mechanical behaviour, often exhibiting brittle behaviour. Fundamental theories of strength and fracture have been developed to explain the mechanical behaviour of brittle solids in general, and a number of standardized strength-testing procedures have been devised. Hence, consideration is given in this Chapter to the application and relevance of these theories to carbon materials under load and describes any required modifications and developments of test procedures. Particular emphasis is given to the influence of structural features, e.g. porosity and pore-wall structure, on the strength properties of carbon materials. It is shown that the factors which determine the mechanical properties of carbon materials are numerous and complex, and that a more detailed understanding of the strength-structure relation for these materials is necessary if materials of specific, optimum mechanical characteristics for a particular application are to be developed.

MECHANICAL PROPERTIES OF COKES AND CARBON COMPOSITES

J.W. Patrick and D.E. Clarke

Carbon Research Group, Loughborough Consultants Ltd., Department of Chemical Engineering, University of Technology, Loughborough, Leicestershire, LE11 3TF, U.K.

1 INTRODUCTION

The mechanical properties of a solid material define the response of the material when it is subjected to an applied load. When a force (load or impact) is applied to a solid, one of two distinct responses are possible: (1) it may deform; or (2) it may fracture. Cokes, carbons and carbon composites, generally exhibit brittle behaviour and, consequently, fracture is the response of greatest significance. Hence, this Chapter concentrates on fracture although some aspects of deformation will also be considered.

The materials considered here are cokes, e.g. metallurgical cokes, and carbon composites, e.g. carbon fibre-reinforced carbons. There are many similarities between these materials, the most obvious being that they both comprise principally carbon, but there are several less obvious similarities, as discussed later; for example, cokes can themselves be considered as composites. However, there are also many differences between these categories of material, reflecting their respective origins and utilization which, for example, has led to the development of some different testing procedures.

Initially, the nature of these carbon materials is described followed by a discussion of their mechanical properties in terms of the different modes of fracture and, where relevant, of deformation. Theoretical consideration of the mechanical properties of brittle materials follows and this leads logically to a discussion of the relevance of material structure. Finally, a summary of the present state of understanding of the mechanical properties of carbon materials and an indication of future directions for research are presented.

2 NATURE OF COKES AND COMPOSITES
2.1 Types of coke

Coke may be produced from several organic precursors, the most important of which are coals and pitches. The most important cokes produced from coal are the "hard" metallurgical cokes in contrast to the "soft" cokes produced at low temperatures for domestic fuel. In addition there are other cokes for certain specialized applications, e.g. ferrocokes and calcined anthracite. Cokes and/or calcined anthracites are also used to produce a wide variety of materials and although the requirements of the coke differ according to the application, a coke of good strength characteristics is generally required in these processes.

One of the functions of metallurgical coke used in the blast furnace is as support to the burden (in the lower regions of the blast furnace, coke is the only solid material present and thereby provides the permeability of the bed necessary for the counter-current passage of the reducing gases through the blast furnace). It is this function which leads to the requirement of high strength. Foundry coke specifications also include high strength and large lump size (80 - 100 mm diameter), one requirement of the coke being to serve as a heat-resistant supporting skeleton, thereby promoting uniform gas distribution during melting. Smokeless fuels for domestic appliances include cokes produced at low temperatures in normal coke ovens, in special processes such as Coalite or Rexco, or in various briquetting processes. In the case of these cokes the material is required to withstand handling and transfer without undue breakage, so the mechanical properties of the materials are again significant.

Cokes are also produced from coal tar and petroleum feedstocks for use as electrode carbon, armature brushes and other special applications. A major use of pitches is as binders, for the production of briquettes, carbon and graphite electrodes, and refractory materials such as carbon bricks, and as an impregnant in, for example, electrode and carbon composite manufacture. Petroleum pitches are carbonized to produce needle and sponge cokes. Needle cokes have a characteristic fibrous texture with long unidirectional "needles" of coke. They have a crystalline structure of high density, low resistivity and low coefficient of thermal expansion and are used in nuclear-grade graphite, aerospace components as well as in conventional graphite applications, e.g. electrodes. Sponge cokes, varying from light honeycomb structures to dense isotropic structures, are used as the solid feedstock for the production of anodes (a pitch binder is the other principal component) for the aluminium industry. In all instances, the mechanical behaviour of the coke is an important property.

2.2 **Influence of production conditions on coke properties**

The properties of coke are influenced primarily by the nature of the parent material being carbonized and for cokes made from coal, the rank of the coal or the average rank of the blend of coals from which the coke is made is of prime importance. It is common practice to blend feedstock coals (sometimes as many as 10) for charging to the coke oven, the object of blending being to produce a consistent coke (in particular, consistently high coke strength) from a wide range of coal types. Coals may be classified, according to the response of their constituent macerals to reflected light, into two classes: reactives and inerts. Reactive macerals melt on heating at about $400\,^{\circ}C$, the inert macerals do not fuse. For optimum coke production there is an optimum ratio of reactives to inerts for each particular coal and coal blend. In many instances, inert materials such as coke breeze or anthracite fines are added to the coal blend to supplement the inert macerals, to increase coke size, strength and yield. These inert materials reduce contraction, and thereby reduce fissuring during the postplastic phase of coking. The size of the inert materials and the coal particles also influences resultant coke properties.

In addition to the nature of the coal charged to the coke oven, carbonization conditions, such as carbonization temperature, charge density, rate of heating, oven design, as well as special conditions such as coal preheating, also influence the mechanical properties of the resultant coke.

The grade of petroleum coke produced from the heavy residual fractions of crude oil is very much dependent on the nature of the feedstock with premium needle coke being produced from highly aromatic thermal tar, pyrolysis tar or decant oil stocks and sponge coke being produced from feedstocks containing a high proportion of resins and asphaltenes.

2.3 **Types of composites**

A composite material may be defined as that which is formed when two or more physically distinct and mechanically separable materials are combined to form a product the properties of which are superior, and possibly unique in some specific respects, to the properties of the individual components.

All of the coke materials discussed so far can in fact be considered to be carbon composites. For example, briquettes comprise coke particles from coal bound together in a matrix by binder coke. Similarly, anodes comprise grist coke bound together in a matrix by binder coke. Metallurgical coke comprises two main components: inert material which has not passed through a fluid phase on carbonization, e.g. inert macerals and mineral matter; and optically anisotropic carbon material ranging in unit size from 1 µm to 100 µm in diameter. Hence, metallurgical coke can also be classed as a composite material when considering mechanical and physical properties. For example, discussions of strength of metallurgical cokes can involve strength of the individual textural components and the strength of bonding at the interface/boundary between components. However, generally, carbon composites are more usually considered to comprise materials that are manufactured from carbon fibres.

There are three main types of composite materials that are prepared using carbon fibres; carbon – polymer/plastic composites (carbon fibre-reinforced plastic, CFRP), carbon – metal composites, and in the present context the most important type, carbon – carbon composites (carbon fibre-reinforced carbon, CFRC).

Carbon fibre composites

Recently, high-performance carbon fibres have been developed with high strength and stiffness and also very low density, These carbon fibres are elastic to failure at normal temperatures and, hence, they are creep-resistant and less-susceptible to fatigue, they are chemically inert except in strongly oxidizing environments, and they have excellent thermophysical characteristics. The physical properties of carbon fibres, in particular high strength and stiffness, are potentially very useful in many applications. Carbon fibres, however, do have some disadvantages compared to other materials: they are brittle, have low impact resistance, low break extension, very small coefficient of linear expansion, a high degree of anisotropy both in the direction of the fibre axis and perpendicular to it, and most significantly, for any practical application the fibres have to be joined and held together in some way to form a structural material. Hence, a matrix is required to hold the carbon fibres, either alone or in structures, together. In this situation the carbon fibre can be thought of as a reinforcing agent for the matrix material.

The amount of reinforcement in a composite is normally given in terms of its "volume fraction" (V_f), i.e. the fraction of the total composite volume taken up by fibres. Ideally, the aim is to incorporate as much reinforcement as possible. It may be observed that the irregular dispersion of fibres may have a significant effect on some properties, e.g. transverse strength and modulus.

With carbon/graphite fibre reinforcements, there are numerous classes of structural materials. The fibres may be randomly distributed chopped fibres in a resin matrix used as moulding compounds; continuous collimated fibres in a resin matrix, as in an unidirectional tape, which may be positioned in accord with engineering design; woven fibre fabrics forming two-directional and three-directional (3D) structures; and honeycomb constructions, important to lightweight aerospace structures.

Carbon fibres are available from rayon, polyacrylonitrile and pitch precursors and each has its own characteristic physical properties that contribute to the final properties of the resultant composite. In addition, the matrix has an important and complex contribution to composite properties. The matrix not only binds the fibres together, but also acts as the medium to transfer external loads on the composite to the fibres. In addition, the matrix has several other important functions: it aligns fibres in important stress directions; it separates fibres, thereby preventing cracks from passing from one fibre to another catastrophically; and it protects fibres from mechanical damage and possibly from high-temperature oxidation.

Although both constituents of a carbon - carbon composite comprise very similar materials, the composite behaviour is very complex and depends on the type of reinforcing carbon fibre, the fibre volume fraction, fibre arrangement, and on carbonaceous matrix parameters such as the matrix structure, which may range from isotropic carbon to graphite, the processing conditions and the thermal treatment of the composite.

Preparation of carbon – carbon composites

Continuous carbon filaments are impregnated with an organic matrix precursor, which may be a pitch, a thermosetting resin or a thermoplastic resin. The impregnated fibres are heated to remove the solvent and to increase the molecular weight of the resin. The fibres are then orientated in the desired arrangement and in the required volume fraction and moulded into prepregs which are pressed further and into laminates and cured. The laminates can then be carbonized in an inert atmosphere at $800^\circ - 1100^\circ C$. The weight loss and the volume shrinkage of the organic carbon matrix produces a porous carbon – carbon composite which may need to be densified. Densification of the carbonized composite is carried out either by chemical vapour deposition (CVD) of carbon from a hydrocarbon gas (such as methane or benzene) or by multiple impregnations and carbonizations with a liquid organic resin or pitch.

Through careful selection of the precursor materials and control of the production conditions, carbon – carbon composites may be prepared that have a number of potential advantages relative to conventional bulk materials: they can be made with high specific strength and stiffness; density is generally low; strength can be high at elevated temperatures; impact and thermal shock resistance (toughness) are good; fatigue and creep strengths are good; oxidation (except at high temperatures) and corrosion resistance are good; and stress – rupture times are improved.

3 MECHANICAL PROPERTIES
3.1 Deformation and elastic properties

A simple method of determining the elastic properties of a material is the tensile test. In this test a bar or rod of material of cross-sectional area A is subjected to an increasing force and measurements are made of the distance between two points on the test piece. The results may be plotted as a graph of load (P) against elongation (dL) of the test piece, as shown in Figure 1. In the first region, AB, with increasing load the specimen stretches proportionally and if the load is removed before point B is attained, the specimen returns to its original length (L). This is termed "elastic deformation". However, this does not apply if point B has been passed. For example, if the load is removed when point C has been attained, the original length is not recovered and the specimen is permanently stretched to length $L + L_c$, <u>i.e.</u> "plastic deformation".

If the load is increased further, the test specimen continues to stretch until point D is attained, which is the maximum load that the specimen can support. From point D onwards, the specimen is unable to support as much load as previously because it has developed a thin spot or "neck" and eventually it breaks, point E. The load at failure/fracture (P_f) is used to calculate the "failure stress" (σ_f) according to the following formula:

$$\sigma_f = P_f / A \qquad (1)$$

The "load - elongation" curve shown in Figure 1 is typical of materials that deform plastically exhibiting "ductile" behaviour (i.e. fracture occurs between points D and E in Figure 1) but some materials do not deform plastically. They are elastic up to the point at which they break (<u>i.e.</u> fracture occurs between points A and B in Figure 1). This behaviour is typical of "brittle" materials such as most cokes and carbons. Hence, ideal brittle fracture occurs without significant plastic flow and after fracture the material returns to its original state except that new fracture faces are created.

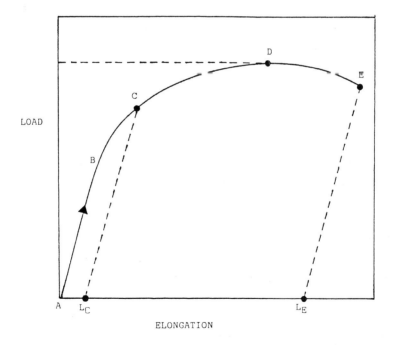

Figure 1   Result of a tensile test. The sample fractures at point E, after having stretched elastically (AB) and plastically (BE).

The values obtained from the tensile test only apply to a specimen of stated dimensions; for example, a material of larger cross-sectional area will support a larger load before failure. Hence, it is necessary to derive a value that is characteristic of the material itself by calculating the amount of force (i.e. stress) required to produce a given degree of stretch or compression (i.e. strain).

Nominal stress (σ) = load (P) per unit area (A):

$$\sigma = P/A \tag{2}$$

Nominal strain (e) = amount of elongation under load per unit length:

$$e = l/L \tag{3}$$

where l = total amount of elongation, and L = original length.

Hooke's Law states that deformation of an elastic body is directly proportional to the load producing it. For small strains, strain is proportional to deformation and stress is proportional to load. Hence, stress is proportional to strain during elastic deformation:

$$\text{stress/strain} = \sigma/e = \text{constant} \tag{4}$$

Young established this constant, Young's modulus (E), to be characteristic of the material tested. Young's modulus is a measure of "stiffness" of a material and fundamentally describes the maximum force required theoretically to either separate the atoms of a material or compress them together.

3.2 Failure

There are several different modes of failure of a material but failure of carbon materials is concerned primarily with: fracture that occurs with little or no plastic deformation, termed "brittle fracture"; fracture that results from a large number of repeated stress cycles (termed "fatigue") at stresses below the yield stress; and abrasion or wear.

3.3 **Brittle fracture**

Fracture is the separation of a material into two or more parts. The nature of the fracture differs with the material and is affected by the type of applied stress, geometrical features of the sample and environmental conditions, e.g. temperature. Brittle fracture occurs by the very rapid propagation of a crack along a grain boundary path (intergranular fracture) or along cleavage planes within the grains (intragranular fracture), after little or no plastic deformation. Theoretical fracture strengths of materials may be calculated from the strength of interatomic or intermolecular bonding. However, these strengths are never attained in any practical application due to the presence of sub-microscopic flaws within the material causing stress concentration under loading. As discussed later, these flaws may be simply dislocations in the structure on the atomic or molecular scale, grain boundary defects, cracks or pores. The stress necessary to cause brittle fracture is inversely related to the size of the defect and essentially strength is determined by the length of the largest defect existing prior to loading.

3.4 **Fatigue**

Fatigue fracture may be defined as the ultimate failure of a material subjected to many cycles of varying stress, the maximum stress applied being less than that required to cause failure if applied continuously. The magnitude of the stress amplitude required to cause fatigue is less when the number of cycles is larger. One common method of representing fatigue is to plot the number of cycles to failure (N) as a function of the stress amplitude (S). The slope of the resultant S - N curve is a measure of the resistance of the material to fatigue and the actual shape is specific to the material tested. Fatigue is rarely of significance for coke materials but is important in many applications for fibre composites, the fatigue resistance of which is dependent on the fibre arrangement, volume fraction of fibre, as well as the fibre and matrix properties.

3.5 **Abrasion and wear**

Abrasion and wear of materials is encountered in any application in which there are moving parts. Abrasive wear is the process whereby a hard rough surface contacts and moves over another surface, so that one or both are subjected to attrition. This effect is of importance to metallurgical coke descending the blast furnace as it may cause a reduction in size. As a consequence, a number of testing procedures have been devised to measure the abradability of coke, as discussed in detail later.

3.6 **Effects of temperature**

Carbon materials in many applications are subjected to elevated temperatures and, consequently, it is important to establish the influence of temperature on the stress – strain relations for these materials. However, mechanical properties are difficult to measure at elevated temperatures and for certain materials, e.g. cokes, only limited studies have been carried out. With resin-based carbon fibre composites, the resin matrix is the strength-limiting factor at high temperatures. Consideration must also be given to differences in the coefficients of thermal expansion for the fibres and the matrix. At elevated temperatures, the fibres and the matrix may expand to different extents generating interfacial shear stresses.

Normally, the strength and stiffness of a material decreases with increase in temperature due to increased thermal agitation of the component atoms and molecules. However, carbon materials are somewhat unusual in that strength and Young's modulus increase with increase in temperature. This behaviour is attributed to a void-filling mechanism, whereby thermal cracks and similar defects formed during the cooling stage of production/processing are closed on reheating.

3.7 **Effects of rate of loading**

The application of a constant load that would not produce failure in a standard tensile test, may cause substantial deformation and eventually even fracture of a material, if the load is applied over a long period of time. This behaviour is termed creep and is particularly significant at elevated temperatures. Although creep may not be a principle consideration for coke materials, it is very important for carbon composites in applications such as turbine blades and aircraft structures. At high temperatures, limited plasticity is possible particularly in polycrystalline carbons, and this may have several origins including dislocation motion, grain-boundary sliding or even softening of minor phases, e.g. thermoplastic matrices in carbon composites. Generally, significant plasticity is not observed and the almost universal consequence is that cracks are eventually nucleated at the boundaries. This can lead to several phenomena such as changes in shape of individual grains or the sliding of adjacent grains relative to each other across a boundary. There are numerous mechanisms whereby grain-boundary sliding may cause grain-boundary cracks; however, regardless of the mechanism of formation the most important consequence is that the presence of such cracks, which often link up along adjacent grain boundaries, may lead to premature brittle fracture.

4   **TEST PROCEDURES**
Coke materials

In almost all applications, the most important mechanical property of a coke is strength. In the coking industry, "strength" is commonly considered to mean the resistance to size degradation. As a consequence, several closely-linked, empirical strength tests, unique to the industry, have developed. The majority of industrial tests involve dynamic loading either in the form of shatter tests or revolving drum tests.

Shatter test

The shatter test is the subject of several standards: ISO Standard 616, ASTM Standard D,3038-72 and British Standard 1016, Part 13.3. The standards differ only in detail and each specification describes a method of test whereby a representative sample of coke (25 kg) is subjected to impact breakage by allowing it to drop four times from a height of 1.83 m (6 ft). The shatter index is reported as the percentage by weight of the original material remaining above a specified sieve size. The ASTM Standard recognizes the effect of the initial size of coke and describes two alternative procedures, for cokes containing more or less than 50% of material -4 in. in size. The British Standard test uses coke sized -2 in. and the ISO Standard specifies a minimum coke size of 50 mm. In these measurements of the resistance of the coke to impact failure, breakage occurs by volumetric failure, i.e. failure at fissures present within the coke prior to testing, and, hence, is governed by the number and concentration of macrofissures present.

Revolving drum tests

Revolving drum tests are widely used by the coking industry as a method of assessing coke strength. Different countries have developed different drum tests but the tests are all similar in that a charge of coke of specified weight and particle size is placed in a horizontal cylinder of stated dimensions, which is then rotated for a specified time at a specified rate.

The degraded coke is removed and sieved, and the strength value is calculated from the percentage coke on certain screen sizes. The drum is equipped with lifters so that the coke is allowed to drop freely within the drum, resulting in a certain amount of shattering action, particularly during the early stages of the test when coke breaks along the planes of the inherent macrofissures, e.g. volumetric failure. In later stages of the test, size reduction occurs as a consequence of abrasion between the coke pieces. This behaviour leads to the production of material having a pronounced bimodal size distribution. The degree and type of dominant breakage mechanism in these tests is controlled by drum type, duration of test and particularly the size and number of lifters.

In Europe and increasingly elsewhere, the MICUM test as proposed by ISO (ISO 556) and the IRSID test, which uses the same apparatus, are dominant. In the US, the ASTM tumbler test (D,3402-75) is standard for use with blast furnace coke. Other standard drum tests are used in Japan and the USSR (Patrick and Wilkinson, 1978).

Although, the strength of coke at high temperatures is of considerable importance, no internationally standardized elevated temperature test has yet been devised. The principal findings of the studies which have been made are that coke strength is reduced and coke abrasion increased when the testing temperature exceeds the carbonization temperature of the coke, the strength of coke measured at ambient temperature is not related to high temperature strength, and annealing of coke between $600°$ and $1000°C$ may result in increased strength. Long coking times improve the high temperature strength of coke.

Interpretation of tests
Although the tests described are long-established, convenient and useful in an industrial situation, there are many problems in attempting to derive fundamental, universal laws from the results because the results are not strictly comparable, as the test procedures differ in many instances. The problem is exacerbated by the degradation of coke in these tests resulting from a mixture of breakage processes occurring during the extremely complex mechanical treatment. Hence, it is difficult to interpret drum indices as indicators of strength in terms of more basic mechanical properties or to relate them to any other coke parameter such as coke structure. Modern

materials science has developed a clear understanding of the mechanical behaviour, in particular strength - structure relations, of a wide-range of materials, based upon well-defined fundamental theories, and there is a need to apply, with modifications as appropriate, the established theory for brittle materials to coke breakage. This requirement has led to the study of other, more basic measurements of coke strength.

Tensile strength of coke

As coke is difficult to machine or cast into the shape required for direct determination of tensile strength, an indirect, diametral compression technique has been used in a series of studies (Patrick and Stacey, 1976 inter alia). In this test, also known as the Brazilian/Carneiro test, an increasing compressive load is applied across the diameter of a prepared disc-shaped piece of coke. Theoretically, breakage occurs along the line of the applied load due to the tensile forces generated perpendicular to it. The tensile strength (σ) is then calculated from the load at breakage (P) and the specimen diameter (D) and thickness (T):

$$\sigma = \frac{2P}{\pi DT} \qquad (5)$$

The test is useful in assessing the fundamental characteristics of unfissured coke on the macroscale and providing tensile strength data that can be correlated with structural features of the coke.

Young's modulus of coke

Young's modulus of elasticity may be obtained from the stress - strain relation derived from so-called static tests in which the stress is applied relatively slowly or from dynamic tests in which the modulus is calculated from the resonance frequency of longitudinal or transverse vibrations, in a coke specimen of given size and shape (Patrick and Wilkinson, 1978).

For an elastic material such as coke, the stress - strain relation is linear and the slope of the line is a measure of the Young's modulus. A relatively simple method of determining the stress - strain curve is to use strain gauges to measure the strain developed in a coke disc during a diametral-compression test. Alternatively, the testing machine may be fitted with an extensometer system based on any one of a variety of transducer elements whereby either the strain may be recorded on a separate XY recorder or the stress – strain curve may be plotted directly on a chart recorder.

There are several experimental advantages in measuring Young's modulus by dynamic methods, particularly for measurements at elevated temperatures. The objective of this technique is to excite and detect resonance in the specimen and then to measure the frequency of this resonance.

The velocity, V, with which a stress wave is propagated through a homogeneous material is related to the effective elastic modulus, E, and the density, p, of the material by the equation:

$$E = pV^2 \qquad (6)$$

This forms the basis of a test in which measurements are made of the resonant frequencies of longitudinal stress waves generated magnetostrictively in small cylindrical or prismatic bar specimens. The coke specimens can be obtained by core-drilling coke lumps and cutting to the desired size.

The resonance conditions in the rod are determined by the length of the rod, L, and the velocity, V, of the stress wave for the fundamental frequency, f:

$$V = 2Lf \qquad (7)$$

Once the stress velocity has been obtained, the Young's modulus may then be calculated.

This method, however, is not appropriate to tests at elevated temperatures; for such tests a method involving flexural or transverse vibrations may be used. The coke specimen in the form of a rectangular prism, may be suspended in the centre of the furnace by two heat-resistant threads, tied close to ends of the specimen, one thread being attached to a vibrator and the other to a crystal pickup detector. The resonance condition is detected by a large and sudden increase in the specimen vibrations detected by the crystal pickup. Young's modulus, E, is calculated using the equation:

$$E = \frac{1.009 m L^3 f^2}{T^3 B} \qquad (8)$$

where m = mass, L, T, and B are the length, thickness and width of the specimen, respectively, and f, is the fundamental frequency. By this method the Young's modulus may be determined at temperatures up to the maximum attainable by the furnace and the support system.

5 COMPOSITE MATERIALS

<u>Young's modulus (tensile) and tensile strength</u>

Composite tensile properties are measured by applying a load parallel to the fibres in a unidirectional composite. Aluminium end-pieces are bonded to each end of the specimen. The width and thickness of the specimen are recorded and the specimen is aligned in the wedge action jaws of a testing machine. An extensometer of suitable gauge length is attached and a tensile load is applied at a constant cross-head speed. The load - deflection curve is recorded with a load between 5 and 15 % of the anticipated ultimate tensile load. The Young's modulus is obtained from the rectilinear portion of the plot and is calculated using the equation:

$$E = \text{stress} / \text{strain}$$

$$= \frac{P_2 - P_1}{W T} \times \frac{L}{L_2 - L_1} \qquad (9)$$

where E = Young's modulus, P_1 = initial load, P_2 = final load, L_1 = length at load P_1, L_2 = length at load P_2, L = gauge length, W = width, and T = thickness.

After the determination of Young's modulus, the specimen is machined to form a "waist". The width (W) and thickness (T) of the waisted portion are measured and recorded and the specimen is aligned in the wedge action jaws of a testing machine. A tensile load (P) is applied at a constant cross-head speed until failure occurs. The ultimate tensile strength (σ_u) is calculated from:

$$\sigma_u = \frac{P}{W T} \qquad (10)$$

Compressive strength

Measurements of compressive strength and modulus are carried out using a test machine as used for tensile testing. The modulus specimen can take the form of a uniform cylinder and for the measurement of compressive strength, the central part of the specimen is reduced in cross section to give a region of uniform stress. When a unidirectional composite is loaded in compression parallel to the fibres, the mode of failure is dependent on the strength of bonding between the fibre and the matrix. If the bond is weak, the fibres and matrix debond at low applied loads, the fibres buckle elastically, and the compressive strength is considerably less than the composite tensile strength. Even if the interfacial bond is strong, the matrix fails in shear and again the compressive strength is less than the tensile strength. This is particularly the case for discontinuous fibre composites, as matrix shear can occur without the necessity for elastic distortion of the filaments. Composites with high volume fractions of fibre, fail by "creasing" of the fibres in compression.

Flexural strength and modulus

Flexural strength and modulus may be measured using composite bars strained in a three-point bend rig in a testing machine. The span - thickness ratio is important in these tests. It has been established that the values for flexural strength and modulus are reduced by the penetration of the supports and loading nose during testing. This effect may be reduced by increasing the span - thickness ratio.

For the determination of flexural strength, the specimen is loaded at a constant rate and the load at which failure (P) occurs is used to calculate the flexural strength (σ_F) according to the following formula:

$$\sigma_F = \frac{3PL}{2WT^2} \qquad (11)$$

where L = distance between supports, W = width, and T = thickness of specimen.

For the determination of flexural modulus, the specimen is loaded at a constant cross-head speed and the deflection (d) for a given applied load is recorded. The Young's modulus is calculated according to the formula:

$$E_F = \frac{L^3}{4WT^3} \times \frac{P}{d} \qquad (12)$$

Interlaminar shear strength

The interlaminar shear strength (σ_S) of composite bars is measured using bars strained in three-point bending using a small span – thickness ratio. The interlaminar shear strength is calculated according to the following formula:

$$\sigma_S = \frac{3P}{4WT} \qquad (13)$$

6 THEORETICAL CONSIDERATIONS
6.1 The Griffith concept

The modern materials science approach to explaining the mechanical behaviour of a brittle solid placed under stress is based upon the classic work of Inglis (1913) and Griffith (1920) who established that strength is controlled by defects or flaws in the material.

Theoretically, in an ideal material such as a perfect crystal, the ultimate strength would be that required to overcome the forces binding the atoms or molecules together. Calculations of interatomic forces indicate that the theoretical strength is always markedly greater than any commonly obtained practical strength. In an initial attempt to explain this phenomenon, Inglis (1913) calculated that any hole or sharp, surface notch in a material causes the stress within it to be increased locally, *i.e.* surface defects cause stress concentration. The increase in stress concentration, which obeys simple mathematical principles, depends primarily on the shape of the hole or notch.

Griffith (1920) proposed that for crack propagation to occur it must be energetically desirable and there must be a molecular mechanism to allow the energy transformation. Griffith attempted to model a crack system in terms of a reversible thermodynamic process by considering the individual energy terms that change as a result of crack extension. It is evident that mechanical energy (strain energy) will decrease with crack extension whereas surface energy will increase as a consequence of intermolecular forces within the material being overcome during the formation of a new fracture

surface. Using a mathematical configuration, termed the Griffith energy-
balance concept, these two factors may be balanced so that a condition is
attained in which the crack is in a state of equilibrium, i.e. will neither
extend nor close, up, and the total free energy of the system is a minimum.
This ability to identify the thermodynamic equilibrium of a crack, using the
laws of energy conservation, is fundamental to predicting the fracture
behaviour of a material.

Having established an energy-balance concept to explain crack propagation,
Griffith extended the theory to consider crack initiation and was able to
describe mathematically the critical condition for fracture in a particular
material under certain constant load, e.g. uniform tension stress:

$$\sigma_L = (2 E y / n c)^{1/2} \quad (14)$$

where σ_L = applied tension normal to the crack plane, E = Young's modulus,
y = free surface energy per unit area, and c = half total crack length.
Application of the theory indicates that if a crack is smaller than a
critical length it consumes more energy than it releases as relaxed strain
energy and, therefore, the conditions are unfavourable for crack
propagation. If a crack is larger than this critical length, these
conditions are reversed and the crack is producing more energy than it is
consuming and, therefore, propagates. The critical length is termed the
"critical Griffith crack length", a value for which is dependent upon the
material and each particular stress within it. An important implication to
brittle fracture is that provided the crack is shorter than the critical
length, the crack will not propagate however large the stress at the tip may
be. Hence, it is important to identify the size order of the critical
Griffith crack length for each material.

The fracture mechanics derived by Griffith have been applied to describe the
strength of a wide-range of brittle materials. With increasing imperfection
of the material structure, grain boundary defects, grain size and porosity
become the source of Griffith flaws responsible for breakage under applied
load. Thus, in low porosity, high density materials with distinct grain
boundaries, grain size and grain boundary defects are the most significant
factors influencing strength, whereas for high porosity materials, pores
being the larger defects, control the critical flaw size.

With increasing porosity, the strength of a material decreases. However, in addition to total porosity, the pore size and shape are also of importance. Only those large pores with a sharp tip of some form and crack-like pores of a length greater than the Griffith critical crack length possibly constitute the effective flaws whereas pores which are blunt, rather than being crack initiators, may act as crack stoppers.

6.2 Interfacial effects
Fibre – matrix interface

The mechanical and physical properties of composite materials are markedly dependent upon the structure and properties of the fibre – matrix interface. The bond between the fibre and the matrix is the load-transfer agent, i.e. the stresses acting on the matrix are transmitted to the fibre across the interface. Composites with weak interfaces have relatively low strength and stiffness, but high resistance to fracture whereas materials with strong interfaces have high strength and stiffness but are very brittle. This effect is related to the ease of debonding and pull-out of fibres from the matrix during crack propagation. The nature of the bond (chemical or physical) is specific to the composite being dependent on the type of fibre and matrix used.

In a simple system, bonding at the interface between a liquid matrix and the carbon fibre during composite fabrication, may be attributed to five main mechanisms either in isolation or in combination: adsorption and wetting, interdiffusion, electrostatic attraction, chemical bonding and mechanical adhesion. There is no completely satisfactory method of measuring the strength of the bond between fibre and the matrix although it is considered important to have some measure of the strength for the evaluation of composite properties and the development of interfaces of appropriate strength characteristics.

There are two main approaches to determinations of bond strength, one involving tests with single fibres and the other concerned with unidirectional laminae.

In the single fibre test, a parallel-sided block of resin containing a short fibre along the central axis is compressed along the axis parallel to the fibre. Shear stresses are created at the ends of the fibre because of the different elastic properties of the fibre and matrix. It is established that the interfacial shear stress is approximately 2.5 times the applied compressive stress. Thus, the shear strength of the interface bond may be obtained by measuring the value of the applied compressive stress at which debonding is first detected at the ends of the fibre. This normally involves visual observations and, hence, this method is inappropriate for opaque matrices. The second approach to measuring interfacial bond strength is to test unidirectional lamellae in such a way that failure occurs in a shear mode parallel to the fibres or in a tensile mode normal to the fibres. One method of achieving this is to use the three-point bend interlaminar shear test as described previously.

It is important to determine the relation between the strength of an interface and the ability of the interface to act as a crack stopper. When a crack approaches an interface orientated perpendicular to it, the crack will try to open the interface by pulling the two sides apart. It has been established that if the strength of the interface is greater than one-fifth of the general cohesive strength of the material, the crack will cross the interface and the material will behave as a normal brittle solid. However, if the strength of the interface is less than approximately one-fifth of the general cohesive strength of the solid, then the interface will be broken and a new crack occurs at right-angles to the original one along the interface and a crack stopper is created.

Interfaces in coke materials
Interfacial effects in coke materials are of less significance than in fibre composite materials, because the principal flaws and defects in cokes are pores and these dominate the mechanical properties. However, in some circumstances, interfacial effects in cokes may be significant and principles relevant to carbon fibre - matrix composites may be usefully applied as coke is similarly derived from two components, one of which is fluid during carbonization and acts as a binder for the other components that are inert.

6.3 **Statistical aspects**
In any experiment that involves fracture of brittle materials there is considerable unavoidable test-to-test variation and a large number of samples must be tested to obtain a reasonable assessment of the performance

of a material. Also, the determined ultimate tensile strength depends on factors such as the volume of material stressed, the shape of the test specimen and the manner of loading, e.g. tension, bending or torsion. To obtain parameters that more accurately define the the tensile strength, the data may be subjected to a statistical analysis one such approach being that adopted by Weibull (1936). The Weibull statistical theory of the strength of materials is an attempt to explain some of the experimental phenomena observed with brittle materials.

On the basis of the flaw concept and according to the weakest-link principle, dispersion of test results is due to the dispersion of flaws. Also, as the volume of the material under stress is increased there is a greater probability of encountering a flaw to promote fracture, and this can account for the observed dependence of the tensile strength on the type of test to which the material is subjected. The empirical approach of Weibull makes no assumptions concerning the nature of the flaws and their distribution, but the result of the analysis is the arbitrary function of an analytical expression for the probability of failure, F:

$$F = 1 - \exp[-V((\sigma - \sigma_u)/\sigma_o)^m] \qquad (15)$$

where σ = applied stress, V = volume of material under stress, and σ_u, σ_o and m are parameters defining the stress. σ_u is the stress level below which no failures are to be expected; σ_o is a material constant relating to some inherent strength; and m is a material constant relating to material homogeneity. The higher the value of m, the Weibull modulus, the more homogeneous is the behaviour of the material and as m approaches infinity the material behaves as if failure was governed by the maximum stress concept.

Although the expression for the probability of failure has no theoretical basis or physical justification, it nevertheless fits a wide range of experimental observations and it has been used to analyse data for coals, graphites and cokes. For cokes, some samples fracture at very low levels of applied stress; hence σ_u may be assumed to be zero without introducing any significant error and the equation may, therefore, be simplified to:

$$1 - F = \exp[-V(\sigma/\sigma_o)^m] \qquad (16)$$

where: (1 − F) is the probability of survival.

A plot of log log $[1/(1 − F)]$ against log σ yields a straight line, the slope of which gives the Weibull modulus, m, i.e. the characteristic homogeneity of the material.

7 STRUCTURAL CONSIDERATIONS
7.1 Influence of porosity in cokes

Metallurgical coke is a high porosity material, the volume porosity regularly being in excess of 50 % and so the dominant structural features are associated with porosity. The porous structure is, however, extremely complex with a wide variety of shapes and sizes of pores and this makes characterization of the structure difficult. The heterogeneity of coke, especially as produced on the large scale compounds this difficulty, but by using modern computerized instrumentation and a statistical approach in conjunction with careful sampling procedures to ensure that a representative sample is examined, it is possible to develop methods whereby the necessary comparative experimental data may be obtained.

Brittle strength is controlled by flaws or defects in the material structure. Consideration of the mechanical properties of cokes in general leads to the conclusion that coke strength is governed primarily by the nature and extent of the porosity and following Griffith theory, it is the larger pores which are likely to have the major influence. Optical microscopy allied to an image analysis system is one of the most appropriate methods of measuring these features. Generally, measurements have been made of the volume porosity, the pore intercept size (a chord sizing based on horizontal intercepts), the pore-wall intercept size (the inter-pore spacing), the number of pores and the caliper diameter (Feret's diameter) at different orientations whereby the largest and smallest pore dimensions are obtained. From these measurements, parameters describing the pore shape can also be derived so that a description of the porous structure is obtained in terms of the pore shape, pore size and inter-pore spacing both as average or extreme values and the distributions of these values.

Comparing the tensile strength of cokes with the volume porosity shows that, as with other brittle materials, the tensile strength generally decreases with increasing porosity. However, there is no simple correlation of wide applicability between the tensile strength and any one structural parameter, indicating a complex strength – structure relation. By combining the effects of several structural parameters, relations between strength and structure that are highly significant statistically have been formulated. The relation:

$$\sigma \times n = c + k\,(w/p) \qquad (17)$$

(where σ = tensile strength, n = number of pores, w = inter-pore spacing, p = pore intercept size, and k and c, constants)

is based on the concept that strength is directly proportional to the amount of solid material available to carry the load and inversely related to the porosity as given by the pore size and the number of pores (Patrick, 1983). Although this correlation has been found widely applicable and useful, it remains largely empirical and can lead to problems as regards interpretation. For example, the equation predicts that to produce a stronger coke at fixed porosity, the porosity is best distributed in the form of a small number of large pores, which is clearly contrary to the accepted theory and practical experience of the strength of materials. This may be overcome by modifying the equation so that:

$$\sigma \times n = k\,(w/p^2) - c \qquad (18)$$

However, although this has been shown to be useful in a predictive sense (Patrick, 1983), it is still empirical and may only be applied with confidence for the range of values used to obtain the correlation coefficient.

To improve the strength – structure relation it is necessary to apply the Griffith theory of brittle fracture which predicts that the flaw size of a material should be related to strength by an inverse-square root relation. The mean pore intercept used in the previous equation is a reasonable comparative measure of pore size when other factors are constant, but as a characterization of the maximum dimension of a porous feature it is inadequate. The maximum Feret diameter (F_{max}) is a more appropriate measure of the maximum dimension of the pores. Also, the minimum Feret diameter (F_{min}) is a reasonable measure of the minimum pore dimension and the ratio of the maximum and minimum diameters provides an assessment of the ellipticity or aspect ratio of the pores, a useful shape factor in terms of the extent of stress concentration effect.

In general F_{max} (Patrick, 1983) relates to tensile strength according to the following inverse-square root relation:

$$\sigma = k \, (F_{max})^{-0.5} \exp[-2 \, (F_{max}/F_{min})^{0.5} \, Py] \tag{19}$$

where σ = tensile strength, F_{max} = maximum Feret diameter, F_{min} = minimum Feret diameter, and Py = fractional volume porosity.

The equation combines parameters measuring the maximum dimension of the pores taken to be critical flaws, a pore shape factor which allows for stress intensity and its dependence on the sharpness of the tip of the flaw, and the porosity which reflects the amount of material able to carry the load. Using this equation, it has been possible to predict coke tensile strength from structural data.

7.2 **Carbon texture in cokes**

Metallurgical and electrode cokes, pitch-based carbon fibres and graphites for nuclear reactors have characteristic microstructures which may give a direct indication of the mechanical properties. The different types of microstructure in cokes can be established by means of polarized light microscopy to assess the optical texture. Several nomenclatures have been devised to distinguish between isotropic and anisotropic cokes and to separate the anisotropic units on the basis of size and shape but all follow a general pattern.

For cokes derived from coal the nature and size of the coke optical texture decreases with decreasing rank of the parent coal. In general, high-rank, low-volatile anthracites and semi-anthracites carbonize without significant softening and the resultant carbon is highly anisotropic being of extensive flow or domain type anisotropy through enhancement of the inherent basic anisotropy of the parent coal. Prime coking coals produce cokes with the largest size of optical texture, ranging from mosaics to flow anisotropy, and are the most complex to analyse. Cokes from low-rank, high-volatile, weakly caking coals contain less anisotropic carbon with the predominant anisotropic texture being of fine-grained mosaics. The low-rank, non-caking bituminous and subbituminous coals produce cokes which appear to be isotropic. In some cokes of this type, however, anisotropic units may be present of size less than 0.3 μm diameter, i.e. beyond the resolving power of the optical microscope, but these may be distinguished using an etching technique followed by examination using scanning electron microscopy.

7.3 **Microstructure and fracture in cokes and composites**

Studies of the tensile strength of coke initially concentrated primarily on the influence of the coke porous structure and assumed that any variations in the textural/structural composition of the pore-wall material had only a secondary influence. However, there is evidence to suggest that different carbon structures as determined by optical microscopy may have different strength characteristics. It has been proposed that cokes of mosaic optical texture exhibit the highest strength as the mosaics of coke are interlocked with each other at the molecular level to form a very firmly bound interface and fissure propagation across the randomly orientated mosaics is not facilitated by this structure. However, it has also been suggested that the boundaries between optical components in coke can act either as strengthening agents, dispersing strain energy and retaining crack energy, or as weakening agents, acting as potential sites for stress-induced fissures depending on the orientation of the structures with respect to the propagating crack.

As discussed previously, the fracture and strength of fibre - matrix composites are related directly to the size of defects and flaws within the structure. These defects may occur within the fibre, the matrix or at the interface between the two and may be observed directly, using optical, scanning electron or transmission microscopy, or inferred from the results of test procedures.

In cokes, it is evident that no single feature of structure can be used to predict the fracture properties of the material accurately. The porosity and porous structure are clearly important factors complicated by the fact that spherical pores can act as crack-stoppers through the redistribution of the stress field around the pore and the absorption of the associated strain energy. Recently, it has been proposed (Patrick and Walker, 1987) that under tensile loading the mechanism of coke failure involves the formation of stable microcracks, initiated at the larger pores, the number and length of which increase with increasing load. Eventually, a flaw of critical size, several millimetres in length and consisting of a series of pores and interconnecting microcracks, is formed and then propagates uncontrollably through the specimen. The role of the nature of the carbon matrix of the coke in the formation of these microcracks is uncertain but it is clear that the structure of the pore-wall material may also influence the strength of coke materials.

7.4 **Fractography**

Scanning electron microscopy with its large depth of focus is well-suited to the examination of rough fracture surfaces and the technique, termed fractography, has been used successfully to view directly the three-dimensional nature of cokes and to infer their mode of failure under stress. The three-dimensional form of the structural units of coke fracture surfaces viewed directly by scanning electron microscopy may be classified on the basis of their appearance into five broad categories termed: flat, lamellar, intermediate, granular and inert. Marked differences in the appearance of fracture surfaces of textural components present in coke are evident using SEM, implying different modes of failure. Granular components (approx. equivalent to mosaic optical textures) appear to fracture in an intergranular mode whereas lamellar components (approx. equivalent to flow anisotropy optical textures), if aligned circumferentially to a pore surface, fail by translamellar fracture, or if aligned radially, cleave similar to mica. Consequently, there are significant differences in the surface roughness of the fractured components, implying a corresponding variation in surface energy which according to Griffith theory of brittle fracture will influence coke strength. To investigate possible relations between the strength and SEM textural composition of cokes, tensile strengths of cokes have been correlated with the proportions of individual components present in the cokes. Preliminary results have indicated that increasing coke tensile strength may be reasonably correlated with increasing content of intermediate components and decreasing proportions of medium and fine granular components (Patrick and Walker, 1985). This type of study is still in the early stages of development and work is currently in progress to establish relations of general applicability between the tensile strengths, textural composition and pore structure of carbon materials.

8 CONCLUSIONS

The factors that determine the mechanical properties of carbon materials, such as fibre - matrix composites and metallurgical cokes, are numerous and complex. No single feature of composite structure can be used to predict mechanical properties accurately. The importance of flaws and stress concentrations they induce in considerations of strength is emphasized in the materials science approach to the strength of brittle materials. The porous structure is clearly an important factor as tensile strengths of coke may be related mathematically to pore structural parameters. The structure of pore-wall material and the interfaces between components may also influence strength characteristics. It is evident that the mechanical behaviour of carbon materials is dependent upon both the structure of pore-wall material and the arrangement of this material in relation to the porosity. If it is possible to understand the strength - structure relations for carbon materials in more detail, then the possibility of producing materials of specific, optimum mechanical characteristics for a particular application may be realized.

References.

The majority of the material concerning the mechanical properties of coke is based upon the work of the authors and their co-workers but limitations of space preclude full acknowledgement of the numerous authors whose work has formed the basis of this review.

Hays, D., Patrick, J.W. and Walker, A. (1982). A scanning electron microscopy study of fractured and etched metallurgical coke surfaces. Fuel 62, 232-236

Inglis, C.E. (1913). Stresses in a plate due to the presence of cracks and sharp corners. Trans. Inst. Naval Arch. 55, 219-230

Griffith, A.A. (1920). The phenomena of rupture and flow in solids. Phil. Trans. Roy. Soc. London 221A, 163-198

Patrick, J.W. (1983). Microscopy of porosity in metallurgical cokes. J. Microsc. 132, 333-343

Patrick, J.W. and Stacey, A.E. (1976). The tensile strength of metallurgical cokes. 4th London Int. Conf. on Carbon and Graphite, Soc. Chem. Ind. 634-643

Patrick, J.W. and Walker, A. (1985). Preliminary studies of the relationship between carbon texture and strength of metallurgical coke. Fuel 64, 136-138

Patrick, J.W. and Walker, A. (1987). A SEM study of the tensile fracture of metallurgical coke. J. Mat. Sci. 22, 3589-3594

Patrick, J.W. and Wilkinson, H.C. (1978). Analysis of metallurgical cokes. Analytical Methods for Coal and Coal Products, Academic Press, New York, Vol. II, 339-370

Weibull, W. (1936). A statistical theory of the strength of materials. Ing. Ventenskaps Acad. Handlinger, Stockholm 151, 1-45; 153, 1-39

Books consulted and recommended for further reading

On coke and coal properties:

Lowrey, H.H. (Ed.) (1963). Chemistry of Coal Utilization, Supplementary Volume, John Wiley and Sons, London

Elliott, M.A. (1981). Chemistry of Coal Utilization, Second Supplementary Volume, John Wiley and Sons, London

On composites and composite properties:

Hull, D. (1981). An Introduction to Composite Materials, Cambridge University Press, Cambridge, UK

Gill, R.M. (1972). Carbon Fibres in Composite Materials, Butterworth & Co Ltd, London

Davidge, R.W. (1979). Mechanical Behaviour of Ceramics, Cambridge University Press, Cambridge, UK

On mechanical properties in general:

Gordon, J.E. (1976). The New Science of Strong Materials, 2nd ed., Penguin Books, UK

Lawn, B.R. and Wilshaw, T.R. (1975). Fracture of Brittle Solids, Cambridge Solid State Science Series, Cambridge University Press, Cambridge, UK

Felbeck, D.K. and Atkins, A.G. (1984). Strength and Fracture of Engineering Solids, Prentice-Hall, Englewood Cliffs, NJ, USA

Chapter 8

The Nature of Coal Material

J.C.Crelling

Dept. of Geology, Southern Illinois University, Carbondale, Illinois 62901, USA.

Summary.

This Chapter is an introduction to the genesis and structure of coals, particularly from the point of view of the geologist. Coal is a rock composed of organic entities, called macerals, with lesser amounts of inorganic minerals, distinct associations of these being called lithotypes. Coal strata occur worldwide and physical structures within these strata (seams) are well defined. Proximate and ultimate analyses of coals broadly characterise the coals and are extended to include calorific values and fluid properties developed on heating. A significant use of coal is for making of metallurgical coke and coals to be carbonized must produce coke of adequate strength and minimum reactivity. Coals and coal blends for coke making are characterised effectively by optical microscopy of polished sections to monitor maceral contents and coal reflectivity which is a rank indicator. The macerals are grouped into Vitrinite, Liptinite and Inertinite for applied petrography. The vitrinite macerals are the most abundant in Northern Hemisphere coals and impart fluidity to coals. The liptinite group are waxy and resinous, derived from spores and cuticles. Inertinite macerals, found more abundantly in Southern Hemisphere coals, have been strongly altered and remain little changed on carbonization. Coal rank, which represents extent of metamorphic change in the dominant coal property, ranges from lignites to anthracites. Rank properties are summarised. Finally, the Chapter discusses such matters as sampling and weathering of coals.

COAL AS A MATERIAL

John C. Crelling

Department of Geology, Southern Illinois University, Carbondale, Illinois 62901, U.S.A.

1 INTRODUCTION

Coal is an extremely heterogeneous material that it is difficult to characterize. Coal is a rock formed by geological processes and is composed of a number of distinct organic entities called macerals and lesser amounts of inorganic substances - minerals. However, coal is not a uniform mixture of these substances. The macerals and minerals occur in distinct associations called lithotypes and each lithotype has a set of physical and chemical properties which also affect coal behaviour.

Each of the various coal macerals have a unique set of physical and chemical properties which control the overall behaviour of coal. Although much is known about the properties of minerals in coal, for example, the crystal chemistry, crystallography, magnetic and electrical properties, surprisingly little is known about the properties of individual coal macerals. Two of the main reasons for this lack of knowledge is that they are extremely difficult to separate from the coal and that they are non-crystalline organic compounds and, therefore, not good subjects to analyze with such standard methods as X-ray diffraction or electron-microprobe analysis. The most successful characterization of coal macerals to date has been by petrographic methods, in which individual macerals do not have to be separated. In the steel industry, for example, petrographic techniques have proven so successful in allowing the prediction of the coking properties of coal that most major steel companies have now established petrographic laboratories. In the last decade, techniques have been developed to separate pure maceral concentrates and the chemical and structural characterization of these materials is now under way.

Coal seams, the basic units in which coal occurs, are composed of layers of coal lithotypes and individual coal seams may also have their own set of physical and chemical properties. For example, even if two coal seams have the same maceral and mineral composition, the seams may have significantly different properties if the lithotypes in the two seams are different.

Because coal seams always occur in association with other strata, the enclosing rocks immediately above and below a coal can also affect the properties of the coal seam. This aspect is of particular importance in mine design, production and strata control.

Thus, the compositional characterization of a coal seam must take place at a number of levels including: 1) the macerals, 2) the lithotypes, 3) the entire seam, and 4) the association of the seam with its enclosing strata.

In addition to compositional factors, coal properties also change with the rank or the degree of coalification of a given sample of coal. Coal is part of a metamorphic series ranging from peat, through lignite, sub-bituminous and bituminous coal to anthracite. The geologic factors of temperature, pressure and time alter the original precursors of coal through this metamorphic series. As the rank of the coal changes, the properties of the coal macerals change progressively and, therefore, so also do the properties of lithotypes and the entire seam.

Because of these factors, coal characterization requires a detailed knowledge of both the maceral composition and rank of coal. All coal properties are ultimately a function of these two factors.

2 OCCURRENCE OF COAL
2.1 Distribution

Although coal particles are scattered throughout many rock units, most coal occurs in seams which can range in thickness from a few millimeters to over seventy meters. In most of the world, anthracite and bituminous rank coals tend to occur in thinner seams of one to two meters while the lower rank coals, lignite and sub-bituminous, commonly occur in thicker seams of up to twenty to thirty meters. While many seams, even minable seams, are of limited area, some seams continuously underlie large areas. The Pittsburgh coal seam extends over 78,000 square kilometers in Pennsylvania, Ohio, West Virginia and Maryland in the United States and it is of minable thickness for 16,000 square kilometers.

Although there may be many named coal seams in a given mining area, relatively few are of the quality, thickness and area to be extensively exploited commercially. For example, while there are hundreds of named coal seams in the U.S.A., over 75% of the coal production comes from less than 25 individual seams.

Coal seams are found in rocks of all geologic ages past the Devonian, although the age distribution is not even. Major coal deposits of the Carboniferous age occur in eastern and central North America and in Britain and Europe. Major deposits of the Permian age occur in the southern hemisphere in South Africa, India, South America and Antartica. In Jurassic times, the major coal accumulation was in Australia and New Zealand and parts of Russia and China. The last great period of coal deposition was at the end of the Cretaceous and beginning of the Tertiary period. Coals originating at this time are found in the Rocky Mountain region of North America, Japan, Australia and New Zealand and parts of Europe and Africa. Because they are younger, these coals tend to be of lower rank, usually sub-bituminous, than the Carboniferous coals. Since this time, some coal has been deposited in scattered locations more or less continuously and this coal tends to be of lignite or brown coal rank.

The distribution of coal seams throughout the world is also not uniform. Most of the world's coal is concentrated in just three countries, the U.S.A., the U.S.S.R. and China. Although the reserve figures vary from source to source, each of these countries have about 25% of the total coal resources, while the rest of the world shares the remaining 25%.

2.2 Coal seam properties

All coals seam properties have a number of structural features including partings, splits, rolls, cutouts, cleat, faults, folds and alterations by igneous intrusions that have a strong effect on the mining and economical recovery of the coal. Partings are layers of rock, usually shale or sandstone, that occur within a coal seam that were caused by an influx of sediment into the original coal swamp. In commercial seams, the partings, of necessity, must be thin (a few centimeters) and few in number. However, even in such seams, the partings can be very extensive. For example, near the base of the Herrin No. 6 Seam in the Illinois Basin, U.S.A., there is a parting known as the blue band which is found throughout the entire basin. When the thickness of a layer of rock within a seam increases to the point where it is no longer practical to mine the parting with the coal, the seam can be considered to have split. While some seams like the Hiawatha in Utah, U.S.A., split into three or more minable seams, the splitting of one seam into multiple seams can present serious problems in mining including correlation and loss of minable thickness.

In addition to partings and splits within the coal seam, the upper and lower surfaces of a seam often pinch and swell to change the coal seam thickness.

Such features have been called pinches, rolls, horsebacks and swells and their occurrence at the base of the seam is usually attributed to differential compaction. Roof rolls are much more common and the protrusion of rock into the coal causes some serious problems with mine roof stability. Cutouts or stream washouts are the extreme case where the protrusion actually eliminates the coal seam. These features are clearly the result of non-deposition or erosion by ancient streams associated with the coal swamps.

Another important feature of coal seams, especially in bituminous coals, is the presence of closely spaced fractures within the coal, called cleat. These fractures are usually perpendicular to the bedding plane of the coal and commonly occur in two sets perpendicular to each other, giving coal the tendency to break into block-like pieces. The cleat controls the ease with which the coal breaks up and has long been used in coal mining. The most prominent cleat is called the face cleat because the working face of a mine is often parallel to this direction. The other cleat is called the butt cleat. Cleats give coal a high permeability to gas and groundwater and also act as sites of mineral deposition. Calcite and pyrite are the most common cleat-filling minerals although other minerals, including gypsum, have been recorded.

Faulting and folding of coal seams by geological forces can also be an important feature. Seams that are faulted and/or folded usually cannot be mined by surface methods and are more expensive to mine than flat-lying seams. When faulting is extensive, even thick seams of high quality may be too discontinuous to mine at all. On the other hand, folding can double the minable thickness of some seams and, thus, increase their value as is the case in the anthracite seams in eastern Pennsylvania, U.S.A.

The alteration of coal seams by igneous intrusions is widespread and is a serious problem in some coalfields such as those in the Rocky Mountain region of the U.S.A. The alteration is thermal and causes an increase in carbon content and a decrease in hydrogen, moisture and volatile matter. In the contact zone, the coal can be transformed into natural coke. Although such contact zones are usually small, extensive areas of natural coke are known and have occasionally been commercially exploited in some areas such as the Raton Mesa in Colorado, U.S.A.

3 **BULK PROPERTIES AND CHEMICAL STRUCTURE**
3.1 **Chemistry**

Although coal is composed of a large number of organic chemical components, there is no true standard organic chemical test for coal. The two major kinds of chemical analyses used are the proximate and ultimate analyses which are standard tests. The proximate analysis consists of a determination of the moisture, ash, volatile matter and a calculation of the fixed carbon value. The moisture is determined by heating the sample at 104° to 110°C to a constant weight. The percent weight loss is reported as the moisture value. The ash in this analysis is the incombustible residue after the coal is burned to a constant weight. It should be noted that the ash value is not a measure of the kinds of or relative amounts of the minerals in the coal. The nature of the mineral matter is discussed below. The volatile matter is a measure of the amount of gas and tar in a coal sample. It is reported as the weight loss minus the moisture after the coal is heated in the absence of air at 950°C for seven minutes. The fixed carbon is not a distinct chemical entity. It is reported as the difference between 100% and the sum of the moisture, ash and volatile matter values. While the proximate value as described above is rather a simple assay of the chemical nature of coal, its value as a quality parameter is well established and is widely used in commerce.

The ultimate analysis consists of direct determination of ash, carbon, hydrogen, nitrogen, sulphur and an indirect determination of the oxygen. The ash is determined as in the proximate analysis and, because all of the values are reported on a moisture-free basis, moisture must also be determined. Typical analyses for coals of different rank are given in Table 1.

Both proximate and ultimate analyses are reported in a number of different ways and care must be taken to be certain of the method of reporting. On the 'as received basis', the results are based on the moisture state of the coal sample as it was received for testing. With the 'dry basis', the results are calculated back to a condition of no moisture and with the 'dry ash free basis', the results are calculated to a condition of no moisture and no ash. These calculations are done so that different coals can be compared on their inherent organic nature. The reporting basis can cause significant changes in the values reported and it is essential that a target value and the value of a sample in question be on the same basis. For example, for a given coal with a moisture of 10%, an ash value of 15%, a volatile matter of 30% and a fixed carbon content of 45%, the volatile matter content would be 33.3% on a dry basis and 40% on dry-ash-free-basis. The corresponding fixed carbon values are 50% and 60%.

Because some components of the minerals such as water from the clays and carbon dioxide from calcite are lost in the high temperature ashing process, the ash value determined is less than the actual mineral matter in the raw coal. A number of corrections for this loss are in use but the one most used in the U.S. is the Parr formula where the corrected mineral matter is equal to (1.08) ash plus (0.55) sulphur. Results reported with this correction are considered to be on a dry-mineral matter-free basis.

3.2 Thermal and fluid properties

The most widely used thermal property of coal is its calorific value which is a measure of the heat produced by combustion of a unit quantity of coal under given conditions. It is usually expressed on a moist, mineral matter-free basis and is used in the classification of coals by rank. The major use of the calorific value is in the evaluation of a coal for use in steam generation. Typical calorific values for coals of different rank are given in Table 1.

The fluid properties of coal are important for the use of coal in coke-making. A coking coal, quite simply, is a coal that when heated in the absence of air will turn into a hard sponge-like mass of nearly pure carbon-coke. Not all coals will produce coke. In general, only coals in the bituminous rank range (from about 0.5% reflectance to about 2.0% reflectance) will react to produce coke, however, not all bituminous coals are coking coals. When a coking coal is heated in the absence of air, it goes through the following stages:

* 350°C - 450°C. Coal softens, melts, turns into fluid, devolatilizes and vesiculates. Usually in this temperature range, the coal develops its maximum fluidity; after this point, the fluidity begins to decrease.
* 450°C - 550°C   The coal begins to harden into coke. In this stage, if the coal is of sufficient rank, liquid crystals begin to form and grow at the expense of the fluid phase. As the coal hardens into coke, these liquid crystals harden into anisotropic domains which give coke its characteristic mosaic cell-wall texture. At this point, the solidified mass is called a semi-coke.
* 500°C - 1000°C There is a further devolatilization which transforms the semi-coke into coke.

In the coking process, the liptinite macerals and most of the vitrinite macerals melt and are reactive. The other macerals, pseudovitrinite and the inertinite macerals, behave in two different ways. The macerals fusinite, macronite, micrinite and sclerotinite, are truly inert in the coking process and are incorporated into the cell-wall structure of the coke much like the way sand and gravel are incorporated into concrete. The other macerals pseudovitrinite, semi-fusinite and semi-micrinite behave in part like reactive macerals and in part like inert macerals.

Some of the most commonly measured fluid properties of coal are the Free Swelling Index (FSI) and the fluidity. The free swelling index is a measure of the free expansion properties. It is determined by classifying the profile of carbon residue (coke button) formed when a sample of pulverized coal is heated in a crucible under given conditions. The coke button is compared to a standard profile chart.

The fluidity test, as in the Gieseler equipment, is a measure of the resistance of a coal sample to a stirrer with a constantly applied torque as the sample is heated through its fluid range at a constant rate. The values usually reported are:
a. <u>Initial Softening Temperature</u> - the temperature at which the coal begins to soften and allows the stirrer to move.
b. <u>Temperature of Maximum Fluidity</u> - the temperature at which the stirrer is moving at its maximum rate.
c. <u>Solidification Temperature</u> - the temperature at which the movement of the stirrer stops as the fluid coal is transformed into coke.
d. <u>Fluid Temperature Range</u> - the temperature range over which the coal is fluid - the difference between the Initial Softening Temperature and the Solidification Temperature.
e. <u>Maximum Fluidity</u> - the maximum rate of movement of the stirrer expressed in dial divisions per minute (DDPM).

Three other tests of the fluid properties of coal are used internationally. The <u>Audibert - Arnu Dilatometer</u> measures the expansion and contraction properties of a coal sample. The coal is placed in a tube sealed with a steel piston and heated at a given rate. The displacement of the piston with temperature is plotted and evaluated. The <u>Roga Test</u> in which the coal sample is mixed with a standard anthracite and coked; the strength of the resulting coke is the value determined. The <u>Gray-King Assay</u> in which the coal sample is coked and the resulting product is compared against standard products.

3.3 **Coking properties**

Originally, charcoal was used in the early blast furnaces and, in some cases, was replaced by anthracite. Eventually, beehive coke (coke made in a beehive shaped brick oven) was used. The beehive process used heat from the pre-heated brick walls and the burning volatile matter to coke the coal. This process caused much pollution and wasted most of the volatile matter by-products. Today, almost all coke is made in the by-product coke oven. This is a rectangular oven which is heated by gas in the side walls. A number of ovens are arranged side-by-side to form a coke-oven battery. A pulverized coal-blend is charged at the top of the oven and then coked in an 18 hour cycle during which the centre

TABLE 1. COAL ANALYSIS
TYPICAL VALUES FOR VARIOUS COAL SEAMS

COAL SEAM:		Illinois 6	Pittsburgh	Buck Mt.	Lower Kittanning	Tioga	Monarch
LOCATION:		Perry Co., IL	Washington Co., PA	Schuykill Co., PA	Somerset Co., PA	Nicholas Co., W. VA	Sheridan Co., WY
Property Analyzed							
*Proximate Analysis	Moisture	8.83	1.64	2.51	0.81	1.92	21.1
	Ash	14.46	6.95	13.59	10.54	15.07	5.1
	Volatile Matter	33.66	35.88	5.63	17.95	30.12	33.8
	Fixed Carbon	43.05	55.53	78.27	70.70	52.89	40.0
*Ultimate Analysis	Carbon	60.46	76.27	76.22	78.28	69.74	55.6
	Hydrogen	4.28	4.76	2.19	4.14	4.18	6.2
	Sulfur	4.45	1.34	0.54	3.07	1.08	0.4
	Nitrogen	1.12	1.03	0.64	1.02	0.95	1.2
	Oxygen	6.40	8.01	4.31	2.13	7.06	31.5
Calorific Value BTU/lb (Moist MM Free)		13159	14890	14419	15508	14761	10190
Free Swelling Index		5	8	0	9	8	0
ASTM Rank		High-Volatile B Bituminous	High Volatile A Bituminous	Anthracite	Low-Volatile Bituminous	High-Volatile A Bituminous	Subbituminous A

*As Received Basis

temperature of the coal charge reaches 1800°F. In this coking process, the coal charge exerts a pressure against the sides of the oven and care must be taken to use only coal blends that will not cause excessive pressures and damage the oven. At the end of the heating cycle, the coke is pushed from the oven and quenched with water.

Coke is charged into a blast furnace with crushed limestone and iron ore (iron oxides). Hot air is blown into the blast furnace near its base resulting in a reaction zone in the furnace in which the iron ore is reduced to metallic iron which drips into the bottom of the furnace and is periodically tapped. The limestone also melts and drips into the bottom of the furnace and floats on top of the metallic iron. It also is periodically tapped. The purpose of the limestone is to act as a fluxing agent and collect the mineral impurities in the furnace. In the blast furnace, the coke remains solid and has three functions:
* It supports the burden of materials that are in the furnace.
* It provides a source of heat for the reactions in the furnace.
* It provides the carbon monoxide and carbon dioxide which chemically reduce the iron ore to metallic iron.

The standard coke quality test used in North America consists of taking a coke sample of a given size and rotating it in a drum at a specified turning rate and number of revolutions. The tumbled coke is then screened and the percentage retained on a 1 inch screen is reported as the coke <u>stability factor</u> and the percentage retained on a $1/4$ inch screen is reported as the <u>hardness factor</u>. For use in modern blast furnaces, a coke stability of 55 or greater is required.

The Micum Tumbler Tests are the international coke strength tests used mainly in Europe. They are rotation tests where coke of a given size is tumbled then screened. The results are reported as:

* <u>Mechanical Strength (Micum M40)</u>
  The percentage of coke retained on a 40 mm screen.
* <u>Abrasion Index (Micum M10)</u>
  The percentage of coke under 10 mm in size.

3.4 Chemical structure

Coal macerals and, indeed, whole coals are composed predominantly of insoluble organic matter which has prevented the use of many powerful analytical techniques for their characterization. A variety of approaches have been used to overcome this problem, including microscopic investigations, pyrolytic degradation, treatment with various chemical reagents and the application of physical methods of analysis. These have served either to break down the organic matter into lower molecular weight and, hence, more analyzable products, or to characterize the whole organic matter without degradation.

Many previous workers have concentrated on the determination of the structure of coal by analyzing the products from the pyrolysis of coal. Although this type of analysis provides rapid results, the products bear an uncertain structural relationship to the organic matter from which they are derived. Consequently, analysis of these products cannot provide structural information on the original materials. Other methods of analysis such as the solid-state techniques of elemental analysis, microscopy, FTIR and ^{13}NMR etc. do provide valuable structural information. However, these techniques can only generate an average or global chemical structure for the material under investigation. From such data, it is very difficult to unravel the structure of each contributing unit and the nature of the bridging elements that hold the whole macromolecule together. For a detailed structural analysis, where individual molecular entities can be identified, the organic matter making up each maceral component of coal must be separated and degraded into simpler, more analytically amenable structures.

However, despite these problems some general information about the overall structure of coal is known; and, because vitrinite is the dominant maceral in most of the coals studied, this general information is really about the chemical structure of vitrinite. The coal structure consists mainly of carbon with hydrogen, oxygen and lesser amounts of nitrogen and sulphur in a three-dimensional cross-linked polymeric network which, in bituminous coal, consists of 2-4 aromatic (benzene) ring units that are cross-linked by aliphatic carbon and oxygen (ether) bridging structures. As the rank of the coal increases above 85% carbon, the structure becomes more ordered, more aromatic and more carbon rich. There is also some evidence to indicate that, within the pore volume of this structure, there is a guest phase consisting of smaller, more aliphatic and mobile substances.

4 COAL COMPOSITION
4.1 The maceral concept

Although the heterogeneous nature of coal has long been recognised in microscopical studies, for example by White and Thiessen (1913, 1920), the term 'macerals' was introduced only in 1935 by Marie C.Stopes. To quote: " The concept behind the word 'macerals' is that the complex of biological units represented by a forest tree which crashed into a watery swamp and there partly decomposed and was macerated in the process of coal formation, did not in that process become uniform throughout but still retains delimited regions optically differing under the microscope, which may or may not have different chemical formulae and properties. These <u>organic</u> units, composing the coal mass I proposed to call <u>macerals,</u> and they are the descriptive equivalent of the

TABLE 2. MACERAL CLASSIFICATION - INTERNATIONAL COMMITTEE FOR COAL PETROLOGY

GROUP MACERAL	VITRINITE	EXINITE	INERTINITE
Maceral	Telinite Collinite Vitrodetrinite	Sporinite Cutinite Resinite Alginite Liptodetrinite	Micrinite Macrinite Semifusinite Fusinite Sclerotinite Inertodetrinite
Submaceral*	Telinite 1 Telinite 2 Telocollinite Gelocollinite Desmocollinite Corpocollinite		Pyrofusinite Degradofusinite
Maceral Variety*	Cordaitotelinite Fungotelinite Xylotelinite Lepidophytotelinite Sigillariotelinite	Tenuisporinite Crassisporinite Microsporinite Macrosporinite	Plectenchyminite Corposclerotinite Pseudocorposclerotinite
Cryptomaceral*	Cryptotelinite Cryptocorpocollinite Cryptogelocollinite Cryptovitrodetrinite	Cryptoexosporinite Cryptointosporinite	

*Incomplete, can be expanded as required.

TABLE 3. MACERAL CLASSIFICATION FOR APPLIED PETROGRAPHY

MACERAL GROUP	MACERAL TERMS ASTM* D2799	ADDITIONAL TERMS USED IN WHITE LIGHT ANALYSES	TERMS USED IN FLUORESCENCE LIGHT ANALYSIS
Vitrinite	Vitrinite	Pseudovitrinite	Fluorescing Vitrinite
Liptinite	Exinite Resinite	Sporinite Cutinite Alginite	Fluorinite Bituminite Exudatinite
Inertinite	Micrinite Semi-Fusinite Fusinite	Semi-Macrinite Macrinite Sclerotinite	

*American Society for Testing and Materials

inorganic units composing rock masses and universally called minerals, and to which petrologists are well accustomed to give distinctive names."

The concept was well received and the term maceral is now recognized around the world. Today, many coal scientisis, especially those outside of North America, regard coal macerals as the smallest microscopically recognizable unit present in a sample. However, in 1958, Spackman presented a concept of macerals that is significantly different and more useful in the chemical aspects of coal science:
'........macerals are organic substances, or optically homogeneous aggregates of organic substances, possessing distinctive physical and chemical properties.....'.

The essence of this concept is that macerals are distinguished by their physical and chemical properties and not necessarily by their petrographic form; thus, even though two substances may be derived from the same kind of plant tissue, for example, cell wall material, and have a similar petrographic appearance, they would be different macerals if they had different chemical or physical properties.

A large number of different macerals have been named and they are classified in various systems. The International Committee for Coal Petrology (ICCP) system is given in part in Table 2. Although this system is useful for some purposes, it is impractical because of the large number of terms to use in routine maceral analysis. For such routine analysis, classification systems with a limited number of terms are needed. In the standard method for maceral use in America, only six terms are required, although some additional terms may be used. Although there is no standard method for the analysis of fluorescent macerals, some additional terms also listed in Table 3 are used for this type of analysis.

4.2 **Vitrinite macerals**

Most vitrinite macerals are derived from the cell wall material (woody tissue) of plants, which are chemically composed of the polymers, cellulose and lignin. Although the details of the vitrinization process are not well understood, it is generally believed that, during diagenesis in the coal forming swamps, the cellulose and lignin in the plant cell walls are chemically altered and broken down into colloidal particles which are later deposited and desiccated. This process, gelification, commonly homogenizes the components so that the resulting macerals are structureless. The variation in the vitrinite macerals is usually thought to be due to differences in the original plant material or to different conditions of alteration at the peat stage or during later coalification.

The vitrinite macerals are the most abundant group and commonly make up 50 to 90% of most North American coals. However, most Gondwanaland coals (coals of the Permian age in the southern hemisphere) and some Western Canadian coals are vitrinite poor. The inertinite macerals dominate in these coals.

The density of the vitrinite macerals varies with coal rank but it is in the range of 1.2-1.8 gm/ml. In recent density gradient separation work, using aqueous cesium chloride, most vitrinite in bituminous coals is found to be in the range of 1.20 to 1.35 gm/ml.

The carbon and hydrogen contents of vitrinite at any given rank are intermediate between those of inertinite and liptinite. Typical values based on chemical analysis of separated maceral fractions from bituminous rank coal range from 4.5-5.5% for hydrogen, 75-85% for carbon, 5-20% for oxygen.

Under the microscope in reflected light, the vitrinite macerals have a reflectance (brightness) between that of the liptinite and inertinite macerals. Because the reflectance of the vitrinite macerals shows a more or less uniform increase with coal rank, reflectance measurements for the determination of rank are always taken exclusively on vitrinite macerals. The reflectance of the vitrinite macerals is also anisotropic so that in most orientations a particle of vitrinite will display two maxima and two minima with complete rotation. In determining a standard reflectance value for a coal, two sets of 50 to 100 measurements of the maximum reflectance are taken. The values are then averaged and compared. For an acceptable determination, the two mean maximums must agree within 2% mean variation.

Some vitrinite macerals show a weak brownish fluorescence when excited by ultra-violet light as do some semi-inert macerals. It is generally believed that this fluorescence is directly related to the chemical reactivity of the maceral.

Two types of vitrinite macerals - vitrinite and pseudovitrinite - are usually distinguished in North American coals. Normal vitrinite is almost always the most abundant maceral present and makes up the ground-mass in which the various liptinite and inertinite macerals are dispersed. It has a uniform gray colour and it is always anisotropic. With ultra-violet excitation, some normal vitrinite will fluoresce. Although there has been some controversy over the term pseudovitrinite since it was introduced by Benedict et al. (1968), most workers who have studied the vitrinite macerals have generally agreed that there are at least two main types. The terms vitrinite A and vitrinite B were introduced by Brown et al. (1964) based on their studies of Australian coals. Taylor (1966), using transmission electron microscopy, observed that vitrinite A appeared to be homogeneous while vitrinite B appeared to be a mixture of vitrinite A and exinite. Stach (1975) introduced the term vitrinite C to correspond to pseudovitrinite and to distinguish it from vitrinite A. The term telecollinite is widely used by European workers in place of vitrinite A and the term desmocollinite is used for vitrinite B. Alpern (1966) divided collinite into homocollinite which is equivalent to vitrinite A and heterocollinite which is equivalent to vitrinite B. The term pseudovitrinite was introducd by Benedict et al. (1968) and further described by Thompson et al. (1974). Kaegi (1985) also discusses the origin and identification of pseudovitrinite. Although the terms are not strictly synonomous, vitrinite A, telecollinite and homocollinite are roughly equivalent to the term pseudovitrinite.

Benedict et al. (1968) distinguished pseudovitrinite from normal vitrinite in the same coal on the basis of its petrographic features of higher reflectance, serrated and wedged shaped fractures, high relief and lack of inclusions. This distinction has proven valid on a great variety of coals and the two macerals are so distinguished in routine analyses in various laboratories. Benedict et al. (1968) were also able to show that pseudovitrinite was not as reactive in coking as normal vitrinite and they related the degree of inertness (reactivity) of the pseudovitrinite in coking directly to the reflectance difference between the two kinds of vitrinite in the same coal.

Recent work by Crelling (1988) shows that vitrinite and pseudovitrinite can be separated from each other on the basis of density using the density gradient centrifugation technique. These results also show that pseudovitrinite and vitrinite are two distinct and separable macerals with measurable differences in their mode of occurrence, petrography, reflectance, thermodynamic properties, density, chemical composition and reactivity.

4.3 Liptinite macerals

The liptinite group of macerals are derived from the waxy and resinous parts of plants such as spores, cuticles and resin. This group generally makes up to five to fifteen percent of most North American coals although it dominates some unusual coals such as cannel and boghead types. In any given coal, the liptinite macerals have the lowest reflectance. The liptinite macerals are the most resistant to alteration or metamorphism in the early stages of coalification and, thus, the reflectance changes are slight up to a rank of medium volatile coal. In this rank range, the reflectances of the liptinite macerals increase rapidly until they match or exceed the reflectance of the vitrinite macerals in the same coal and, thus, essentially disappear.

The liptinite macerals have the lowest density of any maceral group ranging from slightly above 1.0 to around 1.25 gm/ml. In any given coal, the liptinite macerals have the highest hydrogen content and can also have the highest carbon content too. These macerals are largely aliphatic in character although most do have aromatic components. The hydrogen content in these macerals in bituminous coals can range from 7-10%, the carbon content from 75-85% and the oxygen content from 5-18%.

Sporinite is the most common of the liptinite macerals and is derived from the waxy coating of fossil spores and pollen. It generally has the form of a flattened spheroid with upper and lower hemispheres compressed until they come together. The outer surface of the sporinite macerals often show various kinds of ornamentation. In Paleozoic coals, two sizes of spore are common. The smaller ones, usually <100 microns in size, are called microspores and the larger ones, ranging up to several millimeters in diameter, are called megaspores. Cutinite is found as a minor component in most coals and is derived from the waxy outer coating of leaves, roots and stems. It occurs as long stringers which often have one surface which is fairly flat and the other surface which is crenulated. Cutinite usually has a reflectance that is equal to that of sporinite. Occasionally, the stringers of cutinite are distorted. Resinite is also common in most coals and usually occurs as ovoid bodies with a reflectance slightly greater than that of sporinite and cutinite but still less than that of vitrinite. Some of the larger pieces of resinite may appear translucent with an orange colour. In some coals, particularly in those from the western United States, a number of different forms of resinite may be distinguished using fluorescent microscopy. Alginite is derived from fossil algae colonies. It is rare in most coals and is often difficult to distinguish from mineral matter. However, in ultra-violet light, it fluoresces with a brilliant yellow colour and can display a distinctive flower-like appearance.

With the use of fluorescence microscopy, Teichmuller (1974) defined three new macerals-fluorinite, bituminite and exudatinite. Although these macerals have some characteristic features in normal white-light microscopy, they can only be properly identified by their fluorescence properties.

Fluorinite usually occurs as very dark lenses that may show internal reflections. Fluorinite is also commonly associated with cutinite. Fluorinite fluoresces with a very high intensity with a yellow colour. Bituminite is difficult to detect in white-light viewing and is often mistaken as mineral matter. It is common in vitrinite-poor detrital coals. It occurs as stringers and shreads and fluoresces weakly with an orange to brown colour. This material is similar to what other workers call amorphous organic matter (AOM). Exudatinite is a secondary maceral which appears as an oil-like void filling. It has no shape of its own and can usually only be detected in ultra-violet light in which it weakly fluoresces with an orange to brown colour.

4.4 Inertinite macerals

The inertinite macerals are derived from the plant material, usually woody tissue, that has been strongly altered either before or shortly after deposition by forest fire charring or biochemical processes such as composting. These macerals can make up five to forty percent of most North American coals with the higher amounts occurring in Appalachian coals. In southern hemisphere coals, this group commonly is more abundant than vitrinite. The inertinite macerals have the highest reflectance and greatest reflectance range of all the macerals. They are distinguished by their relative reflectances and presence of cell texture. The density of the inertinite macerals is always higher than that of vitrinite and ranges from 1.35-1.60 gm/ml. The inertinite macerals have the lowest hydrogen content of all the macerals along with very high carbon content with values ranging from 2.75-4.25% hydrogen, 73-85% carbon and 13-25% oxygen. Fusinite is found in most coals and has a charcoal-like structure. It is always the highest reflecting maceral present and is distinguished by cell-texture which is commonly broken into small shards and fragments. Semi-fusinite has the cell-texture and general features of fusinite except that it is of lower reflectance. In fact, semi-fusinite has the largest range of reflectance of any of the various coal macerals going from the upper end of the pseudovitrinite range to fusinite. Semi-fusinite is also the most abundant of the inertinite macerals in most coals. Macrinite is a very minor component of most coals and usually occurs as structureless ovoid bodies with the same reflectance as fusinite. Micrinite occurs as very fine granular particles of high reflectance. It is commonly associated with the liptinite macerals and sometimes gives the appearance of actually replacing the liptinite.

5 COAL RANK
5.1 Coalification

A unique and often troublesome feature of coal that distinguishes it from other fuels and bulk commodities is its property of rank. Coal forms a metamorphic series that ranges from peat through lignite or brown coal to sub-bituminous, bituminous and anthracite. Coal rank, then, can be thought of as the position of a given sample of coal in this series or its degree of maturity, metamorphism or coalification. All coal starts out as peat which is then changed into progressively higher ranks of coal. This transformation is generally divided into two phases. The first which occurs in the peat stage is called diagenesis or biochemical coalification. In this phase, most of the plant material making up the peat is biochemically broken down. Specifically, most of the cellulose in the plant material is digested away by bacteria and the lignin in the plant material is transformed into humic acids and then to humic compounds, humins. Some plant material is also thermally altered by partial combustion or biochemical charring.

Still other plant material such as spores and pollen survive the diagenesis stage without much change. After diagenesis is over and the altered peat is buried, geological forces begin to act in the geological or metamorphic phase of coalification. The factors which are considered important in this process are temperature, time and pressure with pressure being the least important.

The majority of the geological evidence indicates that temperature is the major factor in coalification and the temperature range in which most coalification takes place is between 50°C and 150°C. The depth at which these temperatures occur is a function of the natural geothermal gradient (dT/dZ) which ranges from 0.8°C/100 meters to 4°C/100 meters. Therefore, at these two geothermal gradient extremes, the depth at which 150°C would be reached, assuming a surface temperature of 50°C, is 12.5 km (41,000 ft., 7.75 miles) for the lowest gradient and 2.5 km (8,200 ft., 1.55 miles) for the highest.

Changes in coal rank also cause most of the properties of coal to change (Table 4). For example, as rank increases, moisture, volatile matter and ultimate oxygen and hydrogen decrease and fixed and ultimate carbon, calorific value and reflectance increase. All of these measurements and even some combinations of these have been used as measures of coal rank. However, with the exception of reflectance they all suffer from two major drawbacks. First, none of them actually change uniformly across the rank range of coal, and second, they are all bulk properties of coal and, thus, can be significantly affected by changes in maceral composition having nothing to do with rank. For example, a coal with a higher-than-normal content of liptinite macerals can have a higher-than-normal hydrogen content and, therefore, appear to have a lower rank than it actually does based on a rank parameter independent of composition such as reflectance. The reflectance of coal is based on the amount of light reflected from the vitrinite macerals in a coal compared to a glass standard of know refractive index and reflectance. The vitrinite reflectance of coal changes uniformly across most of the coal rank range. However, it is not very sensitive in the lowest rank range (lignite to lower sub-bituminous rank).

5.2 Commercial classifications by rank

The most important classification for commercial purposes in the U.S. is the American Society for Testing and Materials (ASTM) classification by rank. It is the basis on which most of the coal in the U.S. is bought and sold. This classification, ASTM Standard D 388 shown in Table 5, divides coals into four classes, anthracitic, bituminous, sub-bituminous and lignitic, which are further subdivided into thirteen groups on the basis of fixed carbon and volatile matter content, calorific value and agglomerating character. The fixed carbon and volatile matter values are on a dry, mineral matter-free (d.m.m.f.) basis and the calorific values

TABLE 4

CHANGES IN COAL PROPERTIES WITH INCREASING RANK

	Property	Change
1.	Moisture	Decreases
2.	Volatile Matter	Decreases
3.	Fixed Carbon	Increases
4.	Ultimate Carbon	Increases
5.	Ultimate Oxygen	Decreases
6.	Ultimate Hydrogen	Decreases
7.	Calorific Value	Increases
8.	Aromaticity	Increases
9.	Reflectance	Increases

are on a moist, mineral matter-free basis. In this system, coals with 69% or more fixed carbon are classified by fixed carbon content and those with less than 69% fixed carbon content are classified by calorific value. Thus, all lignitic and sub-bituminous coals and the lower rank bituminous coals are classified by their calorific value. It is also important to note that not all coals fit into this system. This is especially true of coals with a high liptinitic maceral content such as cannel and boghead types.

The other classification systems of importance are the international system of the International Organization for Standardization (ISO) and the British National Coal Board. In the international system, coals are divided into two types: hard coals with greater than 5700 k cal / kg (10,260 Btu / lb) and brown coals and lignites with calorific values less than that amount. In the hard coal classification shown in Table 6, the coals are divided into classes, groups and subgroups. The classes are similar to ASTM groups and based on dry, ash-free volatile matter and moist, ash-free calorific value. The classes are numbered 1A, 1B, 2 through 9. The classes are divided into four groups numbered 0 through 4 on the basis of the Free Swelling Index and the Roga Index. These groups are further broken down into six subgroups numbered from 0 through 5 on the basis of their Audibert-Arnu Dilatation number and Gray-King Coke Type. The system is set up in such a way that all coals are classified with a three digit number where the first digit is the class, the second digit is the group and the third digit is the subgroup.

The lignites and brown coals are only divided into classes and groups. The classes, numbered from 10 through 15, are based on ash-free moisture and the groups, based on dry, ash-free tar yield are numbered 00, 10, 20, 30, 40. The

British National Coal Board classification is similar to the international system and is given in Table 7.

6 PROBLEMS PECULIAR TO COAL AS A MATERIAL
6.1 Sampling

There are a number of features of coal as a material that can cause problems in its use. Four of the most serious of these problems are:
* Variations in coal composition and rank.
* Difficulty in sampling.
* Presence of mineral matter.
* Weathering.

The fact that coal is composed of a large number of different entities and that coals show wide variations in rank make it a difficult material to work with on a routine basis. Because the composition and rank of coal coming into a processing or utilization facility can constantly change, these properties need to be constantly monitored to ensure optimal utilization. In addition, these variations and the layered nature of coal seams make reliable coal sampling difficult. Complex and expensive statistical methods of sampling must often be employed to assure a representative sample and bedding and blending systems must also be employed to guarantee a uniform coal feed from a stockpile of coal.

6.2 Mineral matter

In addition to its organic content, all coal contains significant amounts of inorganic mineral matter. The amounts present in a given coal are a major factor in coal quality. While specifications vary widely, most commercial coals must contain less than ten percent mineral matter (ash).

Mineral matter can come into coal both as it is forming in the peat swamp and after it has formed. The actual minerals in coal are usually identified by X-ray diffraction analysis of the low-temperature ash (LTA) of the coal. It is important to note that although minerals can easily be identified with X-ray diffraction techniques, it is not possible to accurately determine the amounts of the various minerals in coal at this time. Although a large number of minerals have been reported to occur in coal, the four most common minerals are: clays, pyrite, calcite and quartz.

The mineral matter in coal is important in a variety of ways. The chemical breakdown of the sulphur bearing mineral pyrite is responsible for much of the air pollution caused by burning coal and also most acid mine drainage. The

TABLE 5. COAL CLASSIFICATION SYSTEM – AMERICAN SOCIETY FOR TESTING AND MATERIALS

Class	Group	Fixed Carbon Limits, percent (Dry, Mineral-Matter-Free Basis)		Volatile Matter Limits, percent (Dry, Mineral-Matter-Free Basis)		Calorific Value Limits, Btu per pound (Moist,[A] Mineral-Matter-Free Basis)		Agglomerating Character
		Equal or Greater Than	Less Than	Greater Than	Equal or Less Than	Equal or Greater Than	Less Than	
I. Anthracitic	1. Meta-anthracite	98	...	...	2	...	...	nonagglomerating
	2. Anthracite	92	98	2	8	...	...	
	3. Semianthracite[C]	86	92	8	14	...	...	
II. Bituminous	1. Low volatile bituminous coal	78	86	14	22	...	...	commonly agglomerating[E]
	2. Medium volatile bituminous coal	69	78	22	31	...	...	
	3. High volatile A bituminous coal	...	69	31	...	14 000[D]	...	
	4. High volatile B bituminous coal	...	...	...	...	13 000[D]	14 000	
	5. High volatile C bituminous coal	...	...	...	...	11 500	13 000	
						10 500	11 500	agglomerating
III. Subbituminous	1. Subbituminous A coal	...	...	...	...	10 500	11 500	nonagglomerating
	2. Subbituminous B coal	...	...	...	...	9 500	10 500	
	3. Subbituminous C coal	...	...	...	...	8 300	9 500	
IV. Lignitic	1. Lignite A	...	...	...	...	6 300	8 300	
	2. Lignite B	...	...	...	...	...	6 300	

[A] This classification does not include a few coals, principally nonbanded varieties, which have unusual physical and chemical properties and which come within the limits of fixed carbon or calorific value of the high-volatile bituminous and subbituminous ranks. All of these coals either contain less than 48 % dry, mineral-matter-free fixed carbon or have more than 15 500 moist, mineral-matter-free British thermal units per pound.
[B] Moist refers to coal containing its natural inherent moisture but not including visible water on the surface of the coal.
[C] If agglomerating, classify in low-volatile group of the bituminous class.
[D] Coals having 69 % or more fixed carbon on the dry, mineral-matter-free basis shall be classified according to fixed carbon, regardless of calorific value.
[E] It is recognized that there may be nonagglomerating varieties in these groups of the bituminous class, and that there are notable exceptions in high volatile C bituminous group.

TABLE 6. COAL CLASSIFICATION SYSTEM - INTERNATIONAL

GROUPS (determined by caking properties)			CODE NUMBERS										SUB GROUPS (determined by coking properties)		
GROUP NUMBER	ALTERNATIVE GROUP PARAMETERS		The first figure of the code number indicates the class of the coal, determined by volatile matter content up to 35% V.M. and by calorific parameter above 35% V.M. The second figure indicates the group of coal, determined by caking properties. The third figure indicates the subgroup, determined by coking properties.										SUBGROUP NUMBER	ALTERNATIVE SUBGROUP PARAMETERS	
	Free-Swelling Index	Roga index												Dilatometer	Gray-King
3	>4	>45				435	535	635					5	>140	>G_9
					334	434	534	634					4	>50-140	G_5-G_8
					333	433	533	633	733				3	>0-50	G_1-G_4
					332 a / 332 b	432	532	632	732	832			2	≤0	E-G
2	2½-4	>20-45			323	423	523	623	723	823			3	>0-50	G_1-G_4
					322	422	522	622	722	822			2	≤0	E-G
					321	421	521	621	721	821			1	Contraction only	B-D
1	1-2	>5-20		212	312	412	512	612	712	812			2	≤0	E-G
				211	311	411	511	611	711	811			1	Contraction only	B-D
0	0-½	0-5	100 A / 100 B	200	300	400	500	600	700	800	900		0	Non-softening	A
	CLASS NUMBER		0	1	2	3	4	5	6	7	8	9			
CLASS PARAMETERS	Volatile Matter (dry, ash-free)		0-3	>3-10	>10-14	>14-20	>20-28	>28-33	>33	>33	>33	>33			
	Calorific Parameters[a]		-	>3-6.5 / >6.5-10	-	-	-	-	>13950	>12960-13950	>10980-12960	>10260-10980			

CLASSES
(Determined by volatile matter up to 35% V.M. and by calorific parameter above 33% V.M.)

TABLE 7. COAL CLASSIFICATION SYSTEM – BRITISH – NCB

Group	Class	Volatile matter (% dmmf)	Gray-King coke type	Description
100	101	6.1*	A	Anthracite
	102	6.1-9.0*	A	Anthracite
200	201	9.1-13.5	A-G	Dry steam coals
	201a	9.1-11.5	A-B	
	201b	11.6-13.5	B-C	
	202	13.6-15.0	B-G	
	203	15.1-17.0	B-G4	Coking steam coals
	204	17.1-19.5	G1-G8	
	206	9.1-19.5	A-B** / A-D***	Heat-altered low-volatile bituminous coals
300	301	19.6-32.0		
	301a	19.6-27.5	G4	Prime coking coals
	301b	27.6-32.0		
	305	19.6-32.0	G-G3	
	306	19.6-32.0	A-B	Heat-altered medium-volatile bituminous coals
400	401	32.1-36.0	G9	Very strongly caking coals
	402	>36.0		
500	501	32.1-36.0	G5-G8	Strongly caking coals
	502	>36.0		
600	601	32.1-36.0	G1-G4	Medium caking coals
	602	>36.0		
700	701	32.1-36.0	B-G	Weakly caking coals
	702	>36.0		
800	801	32.1-36.0	C-D	Very weakly caking coals
	802	>36.0		
900	901	32.1-36.0	A-B	Noncaking coals
	902	>36.0		

*To distinguish between classes 101 and 102, it is sometimes more convenient to use a hydrogen content of 3.35% instead of 6.1% volatile matter.
***For volatile matter contents between 9.1 and 15.0%.
****For volatile matter contents between 15.1 and 19.5%.

slagging properties of the coal mineral matter can affect boiler fouling and the production of fly ash. The mineral matter can also hinder coal cleaning, coke production and the conversion of coal into other forms of fuel.

6.3 Weathering

Weathered coal is coal that has been altered by natural geological weathering processes. For example, coal is physically broken up as a result of freezing and thawing and is chemically broken down by solution, hydration, hydrolysis, oxidation and various biological processes. These powerful processes are the same ones that reduce rock to soil. Because these processes require water and occur most readily near the surface of the earth, coal seams nearest the surface stand the greatest chance of being weathered. Thus, deep-mined coal is rarely weathered, whereas most outcrop and much of the surface-mined coal has undergone significant amounts of weathering.

Weathering can cause coal to fracture and break down into fine particles and this decrease in size can make the coal more difficult to manage in coal handling systems. It can also lead to problems in the control of bulk density of a coal charge where the oil spray added as a lubricant is lost in the weathering induced fractures. Weathered coal is difficult to clean and has an often serious reduction in calorific value and fluid properties. This latter change can destroy the coking ability of a coal.

Recent studies by Goodarzi and Murchison (1976), Gray and others (1976), Crelling and others (1979) have shown that the presence of weathering in a coal blend used for coking causes the following detrimental effects:
* A loss of coke strength. Even small amounts (<10%) of weathered coal can reduce coke strength.
* An increase in coke reactivity. (It is believed by some workers that coke reactivity should be kept low and uniform).
* An increase in coke breeze (<$1/2$" coke).
* A decrease in coking rate.

Because all of these changes are detrimental to coke quality, the presence of weathered coal in mixes adversely affects the economics of coke-plant and blast-furnace operations for the following reasons: decreased coke stability brings about a corresponding decrease in hot-metal production in the blast-furnace, decreased coking rate can limit coke production or increase coke-oven underfiring requirements and can contribute to excessive pushing emissions resulting from the production of green coke, increased coke breeze means a loss in coke production, with accompanying increased costs. Thus, the use of weathered coal for coke-making should be avoided.

References

Alpern, B, (1966). Un example interessant de houillification dans le bassin Lorrain et ses prolongements. Adv. Org. Geochem. 1964, Pergamon Press, Oxford, p. 129-145.

Benedict, L.G., Thompson, R.R., Shigo III, J.J. and Aikman, R.P., (1968). Pseudovitrinite in Appalachian coking coals. Fuel **47**, p. 125-143.

Brown, H.R., Cook, A.C. and Taylor, G.H., (1964), Variations in the properties of vitrinite in isometamorphic coal. Fuel **43**, p. 111-124.

Crelling, J.C., (1988). Separation and characterization of coal macerals including pseudovitrinite. Ironmaking Proceedings **47**, p. 351-356.

Crelling, J.C., Schrader, R.H. and Benedict, L.G., (1979), Effects of weathered coal on coking properties and coke strength. Fuel **58**, No. 7, p. 542-546.

Goodarzi, F. and Murchison, D.G., (1976). Petrography and anisotropy of carbonized preoxidized coals. Fuel **55**, No. 41, p. 141-147.

Gray, R.J., Rhoades, A.H. and King, D.T., (1976). Detection of oxidized coal and the effects of oxidation on the technological properties. Trans. S.M.E. **260**, p. 334-341.

Kaegi, D.D., (1985). On the identification and origin of pseudovitrinite. Int. Jour. Coal Geol. **4**, p. 309-319.

Spackman, W., (1958). The maceral concept and the study of modern environments as a means of understanding the nature of coal. Trans. New York Acad. Sci. Ser. II **20**, No. 5, p. 411-423.

Stach, E., (1975). E. Stach, G.H. Taylor, M.-Th. Mackowsky, D. Chandra, M. Teichmuller and R. Teichmuller, Stach's Textbook of Coal Petrology 2nd ed. Borntraeger, Berlin and Stuttgart, 1975, p. 59.

Stopes, M.C., (1935). On the Petrology of Banded Bituminous Coals. Fuel **14**, p. 4-13.

Taylor, G.H., (1966), The electron microscopy of vitrinites: in R.F. Gould, ed., Coal Science: Adv. in Chem. Ser. 55, Am. Chem. Soc., p. 274-283.

Teichmuller, M., (1974), Uber neue Macerale der Liptinite-Gruppe und die Entstehung van Micrinit: <u>Fortsch. Geol. Rheinld. u. West</u> Vol. **24**, p. 37-64.

Thiessen, R., (1920). <u>U.S. Bur. Mines Bull</u>. No. **117**, 296 pages.

Thompson, R.R. and Benedict, L.G., (1974). Vitrinite reflectance as an indicator of coal metamorphism for coke making. Carbonaceous Materials as Indicators of Metamorphism, ed. R.R. Dutcher, P.A. Hacquebard, J.M. Schopf and J.A. Simon, Soc. Am. Spec. Paper 153, p. 95-108.

White, D. and Thiessen, R., (1913). <u>U.S. Bur. Mines Bull</u>. No. **38**, 390 pages.

Chapter 9

Coal to Coke Conversion

R.J.Gray

Ralph Gray Services, 303 Drexel Drive, Monroeville, PA 15146, U.S.A.

Summary.

This Chapter is principally concerned with the major topics associated with the conversion of coal to coke. Only a limited range of coal rank i.e. the bituminous coals, produce acceptable metallurgical cokes. The history of coke making is briefly reviewed followed by description of a coke battery and the by-products of coking. Theories of carbonization are abundant but current knowledge of liquid crystal theory has done much to expand the understanding of the origin of cokes and their properties. Coal Rank, Type and Grade are properties used in coal classification for coke making. The coking of a single coal particle and the development of the coke cenosphere are described. The importance of the plastic (fluid) layer, developed within the coal charge in the slot coke oven, is high-lighted including the thermal transformation of coal macerals to coke components. In industrial practice, the ability to predict coke strength or mechanical stability from properties of coal blends is of obvious importance. Much has been done to analyse coke structure in terms of optical microscopy of polished sections, i.e. coke petrography. This exhibits the various isotropic and anisotropic components of coke structure. Cokes from different coal blends have different contents of anisotropic shapes, 0.5 to >25.0 μm in size, arising from the heterogeneous nature of the plastic (fluid) phase of carbonization from which the liquid crystal structures develop. Classifications of anisotropy are listed. The Chapter continues with descriptions of porosity and pore wall structures in cokes. Carbonization variables relate to final coke property. The behaviour of coke in the blast furnace, including discussions of reactivity, aspects of formed coke, pre-heating and co-carbonization conclude the Chapter.

COAL TO COKE CONVERSION

Ralph J. Gray
Consultant - Coal, Coke, Carbons, Ralph Gray Services, 303 Drexel Drive, Monroeville, PA 15146, U.S.A.

1 INTRODUCTION

Coke is a black to silver-grey porous carbon residue from the destructive distillation of organic matter, particularly coal and petroleum residuals. It is generally manufactured in by-product recovery ovens for use in the iron blast furnace, cupola and various metal smelters. It is also used in the chemical industry, in non-ferrous smelters and in the form of specialty carbons such as carbon brick, electrodes, etc. Blast furnace and smelter cokes are usually made from coal while electrode carbons and specialty carbons are commonly made from petroleum residuals. This Chapter deals principally with blast furnace cokes. Only a limited range of coals produce acceptable commercial cokes. These coals are bituminous. Even when coals qualify by 'Rank', they are restricted by ash and sulphur or 'Grade' and by reactivities and inerts or 'Type' for use in coke making. It is rare to find a coal that meets all of the specifications for coke making. Therefore, blending of coals is resorted to and it is necessary to understand the properties of coals and their functions in the coal-to-coke conversion if the product is to meet specified quality requirements.

This Chapter describes the coal-to-coke transformation with particular emphasis on the role of coal macerals and their degree of maturity. The microscopy of coal and coke is emphasized and the effects of operating variables on coke microstructure and macrostructure are discussed.

1.1 History of coke making

Coke was an article of commerce among the Chinese over 2000 years ago. It was used in the arts and for domestic purposes but not in great quantities. Coking is undoubtedly one of the oldest coal conversion processes and it developed with very little dependence on the understanding of the mechanism of coal to coke transformation. Kirov and Stephens (1967), review 'The Early

History of Coal Carbonization'. The earliest record of coal carbonization dates to about 1590 when John Tornborough, Dean of York, was issued with a patent to purify coal and drive-off offensive smells by coking. In 1620, Sir William St. John was granted a patent for a bee-hive oven. Sir John Winters charred coal in 1606 to drive-off sulphur and arsenic. In 1700, J. Becher, a German chemist, patented a process for saving the tar from coking coal. The earliest efforts to carbonize coal were to make it less objectionable for domestic heating. It was later that coking to produce a new fuel was attempted and the interest in coke by-products came even later. For many years, coke was a relatively undesirable by-product of the gas industry.

The first rectangular or retort coke ovens were constructed in 1830 in Germany and, in 1835, retort coke was produced by William Firmstone in Pennsylvania. The greatest growth in the coking industry developed when coke replaced charcoal as the fuel for the blast furnace. In the early years of the industrial revolution, blast furnaces were short stacked and used charcoal as the preferred fuel for sponge iron production. Coal and coke were considered unsuitable for smelting because excessive carbon and sulphur were picked up by the iron. It was competition for wood for fuel, domestic building, ship building, etc., which led to the use of coke from coal in the blast furnace. In 1735, Abraham Darby designed a tall stack furnace to accommodate coke to produce iron. In 1787, Cort and Onions discovered hot-metal puddling to burn-out carbon from pig iron to produce an acceptable iron product. Finney and Mitchell (1961), give an excellent history of the coking industry in the United States. In the U.S.A., anthracite was used as blast furnace fuel as late as 1923.

As coal began to replace wood as the raw material in the charcoal pile, it was natural that coke production began in piles or mounds with flues that were similar to those used for charcoal production. The Making, Shaping and Treating of Steel (McGannon 1970), gives a history of the development of coke ovens.

The bee-hive oven came into use as the demand for coke increased and the sites for coke production became more permanent. The oven consists of a circular, dome-like chamber (bee-hive) lined with firebrick and a flat-tiled floor. Coal was charged through a top opening and a door in the side was used to control combustion air and for levelling and raking the coke from the oven. The bee-hives continued in use until the nineteen sixties in the U.S.A. The bee-hive coke oven gave low coke yields i.e. 50 wt.%, and wasted many of the by-products. Late in the 19th century, coke ovens were developed in Europe to produce coke and recover by-products. These ovens recovered many by-products, one of which was ammonia, for use in fertilizers. Fertilizer became an important coke oven by-product wherever soils were depleted. Eventually, a large chemical industry was developed, based on coal by-products. The by-product industry grew slowly in the U.S.A. until World War I produced a demand for coal by-products for making explosives.

1.2 **The by-product battery**

The coke battery is the basic unit in a coke works. It is a rectangular steel and brick structure. A battery may be over a hundred feet long and contain up to a hundred or more rectangular ovens. Ovens are commonly about 13 to over 20 feet high and up to about 50 feet long and 14 to 24 inches wide (Figure 1). Ovens taper 2 to 4 inches in width to aid in discharging. Each oven holds about 15 to over 30 tons of coal per charge. The oven chambers are lined with silica brick, which has been fired to cristobalite which melts at 1713°C. Each oven is separated from its neighbour by walls, with flues in which gas is burned for heating the coke oven. The entire oven mass is insulated and it sits on a regenerator chamber which sits on a concrete pad (Figure 1). The mass of brick in the battery is held together with steel tie rods and buckstays.

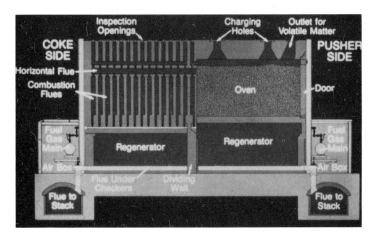

Figure 1. Side View of a By-Product Recovery Coke Oven.

Both ends of the oven are closed with doors. Each oven has two to four lidded openings on top for charging coal. Standpipes at the ends of ovens conduct by-products into collecting mains.

Coal for coking is stored in a bunker at the end of a battery. Coal from the bunker is discharged into a larry car which charges the ovens from hoppers. The pusher machine has a bar for levelling the coal. The area above the levelled coal (tunnel head) conducts gas out of the oven and into the collector mains.

Coking takes about 12 to 18 hours for blast furnace coke but may exceed 30 hours for foundry coke. When coking is completed, the doors are removed and the coke is pushed out into a quench car. Coke is water or dry quenched.

1.3 By-products of coking

Coal decomposes when heated in by-product ovens to form a solid carbon or coke and volatile compounds or by-products. The simple volatile compounds that are produced are called primary decomposition products. These products undergo continued exposure to high temperature as they move through the coke mass and into the collecting system. Thus, most of the by-products are products of secondary reactions. The top of the standpipe which connects the oven to the mains contains a gooseneck with flushing liquor sprays to cool the gases entering the mains from about 1300°C. A ton of coal produces 10 to 16 gallons of tar and light oil, 4.5 to 8.0 pounds of ammonia, 9,000 to 12,000 cubic feet of fixed gas and about 65 to 80 wt.% of coke plus breeze. Breeze is fine ($-^3/_4$ inch) coke. Little or no ammonia is produced in low temperature carbonization but the yield of oil plus tar is about double that of high temperature carbonization.

About half of the nitrogen and all of the phosphorus in coals is retained in coke. Sulphur in coal is partitioned with 50-65 wt.% remaining in the coke.

2 THEORIES OF CARBONIZATION

2.1 Solvent extraction theory is one of the oldest and most widely studied approaches to explain the composition and coking characteristics of coal. According to van Krevelen and Schuyer (1957), Fremy (1861) and Marcilly (1862) conducted systematic solvent extraction studies as early as 1860. In later experiments in coal extraction, beginning in 1910 and continuing for 25 years, pyridine, chloroform, light petroleum, ethyl-ether and acetone were used in a soxhlet extraction apparatus to isolate compounds from coal.

Fischer and Gluud (1916) used benzene for extraction under pressure to obtain "Restkohle", which was insoluble and non-coking and a soluble coking fraction called "Bitumen". These workers also found that some "Restkohle" coked and some did not when the extract was added and the material carbonized.

The Solvent Extraction Theory, in modified form, is currently discussed (see Chapter 2.)

2.2 Transient fusion theory. In the transient fusion theory, coal is likened to $KClO_3$ which melts when heated to form an unstable liquid which decomposes to form a solid. This theory is not seriously considered.

2.3 **Precursors of the metaplast theory**

It has not been resolved whether coals contain a coking material (bitumen) of if the material which causes coal to soften (metaplast) is a product of heating. In addition, does it act as a dispersion agent or a lubricant? van Krevelen and Schuyer (1957) and Kirov and Stephens (1967) discussed coal composition and carbonization theories.

Kreulen (1948) proposed a dispersion micelle theory in which coal is an organosol consisting of an oil dispersion medium (oily phase) and a dispersed phase (micelle phase), which are bitumens. Micelle consists of an oleophilic part (protective body) and an oleophobic part (micelle nucleus) which are humins. According to Kreulen, coking is a process shared by the whole system.

The Lubricant Theory was favoured by many researchers. In the turbostratic lamellar model, coal was considered as consisting of flat molecules that are polycondensed aromatics (lamellae) whose size increases with rank. X-ray studies considered a variety of scattering angles, three types of structure being distinguished. Open Structure <85% carbon, Liquid Structure 85 to 91% carbon and Anthracite Structure >91wt% carbon. The idea that the differentiation between dispersed phase and dispersing medium in coal is one of size and relative mobility was restated. Coal was considered as an isogel that becomes mobile above the gel point and remains solid below the gel point.

Thermobitumen Theory of A.P. Oele et al (1951) postulated that plastic behaviour of coal is due mainly to a liquid primary product formed by pyrolysis. This idea predates the metaplast concept and differs principally in the fact that a thermobitumen implies extensive decomposition.

2.4 **Metaplast theory**.
van Krevelen (1950) studied the kinetics of degasification and divided pyrolytic decomposition into three types of reactions. Additional work by Fitzgerald (1956) and Chermin and van Krevelen (1957) resulted in distinguishing the following three reactions:

I	Coking Coal (P)	$\xrightarrow{K_1}$	metaplast (M)
II	Metaplast (M)	$\xrightarrow{K_2}$	semicoke (R) + primary gas (G_1)
III	Semicoke (R)	$\xrightarrow{K_3}$	coke (S) + secondary gas (G_2)

The metaplast is a pyrolysis product which plasticizes the coal prior to decomposition. The first reaction is depolymerization, followed by cracking and recondensation and then secondary degasification.

2.5 Liquid crystal theory. When the reactive macerals (vitrinite, etc.) in coal are heated, they first lose their bireflectance and become isotropic (characteristic of liquids). The pyrolysis product or metaplast peptizes the coal mass, producing a plasticity which passes through a liquid crystal state (mesophase) that forms an ordered anisotropic mosaic structure in forming the semicoke. This semicoke is further devolatilized and shrinks to form the final coke. Mesophase has been recognized in heated pitches from petroleum residuals and coal tar. Brooks and Taylor (1968) were the first to note the occurrence of mesophase in heated coals. Ramdohr (1928) described differences in anisotropic microstructures in coke compared with natural graphite. Marsh et al. (1982, 1986, 1988) have done much to expand our knowledge of liquid crystal transformations in carbonaceous materials. The development of many speciality carbons, including carbon fibres, (Chapters 1,6) utilizes the liquid crystal concept in the selection of raw materials and the processing to produce new carbon forms with specific textures, forms and strengths (See Chapter 2).

2.6 Liquefaction theory. Neavel (1976) accepts the coal model of Given (1960) which consists of aligned molecular units of variable structures typified by condensed ring systems connected by bridging atoms. The functional groups = O, - COOH, - OH, - C_nH_m are attached to ring carbons and some rings may be saturated with hydrogen.

Neavel (1976) noted that only the macerals exinite and vitrinite become plastic in bituminous coals. He stated that vitrinite, which is the dominant coal maceral, is made up of packets (micelles) or more-or-less aligned molecular units (lamellae) of variable structure typified by a condensed ring system connected by bridging atoms.

Neavel (1976) and Marsh and Neavel (1980), together with Wiser (1968) recognized the relationship between coal liquefaction and plasticity of coking coals. Neavel (1976) proposed "that the development of plasticity in coals is, essentially, a hydrogen-donor liquefaction process in which the solvating and hydrogen-donating 'vehicle' is supplied by the coal itself and it is the progressive reduction of transferable hydrogen inventory which leads to progressive decay of fluidity even under isothermal conditions". Anything that changes the quantity and quality of the vehicle changes the plastic properties of heated coal. Coal rank, type and processing conditions as well as impurities will change the plastic properties of coal. Pressure and large particles retard vehicle loss and trapped vapours enhance plasticity, while oxidation retards liquid yield and low heating of marginally plastic coals destroys plasticity.

3 ORIGIN OF COAL

A brief review of coal origin is helpful in understanding why some coals coke and others do not. It is generally established that coals which range from lignite to anthracite are the products of progressive change of plant materials which derived their energy frcm the sun. Plants, through photosynthesis, change carbon dioxide and other materials into cellulose and protein. In this synthesis, the combustion process is partially reversed and carbon is partly refined back to its condition of high energy potential. The preserved plant materials when buried by sediments were compacted to form low rank coal. Sedimentary build-up consolidates the vegetable debris and squeezes out water, forming low rank coals. As coals are buried deeper, the increased temperature with depth causes devolatilization and formation of bituminous and higher rank coals. Stach et al. (1982) gives a detailed description of coal formation.

In general, it is only the bituminous coals that coke. Coals of lower rank decompose when heated and coals of higher rank do not soften.

4 COAL CLASSIFICATION

Coal is a solid fossil fuel derived from plants that have been accumulated and variously acted upon by geologic processes. It consists of macerals (the organic equivalents of minerals), moisture and mineral matter. Coals are complex, heterogeneous rocks that vary widely in their properties and uses. Their properties can be characterized in terms of Rank, Type and Grade. Rank is maturity. Type is determined by the amounts and kinds of vegetable materials and their state of preservation. Grade is generally regarded as the purity. Most classifications of coal employ a rank parameter such as volatile matter, fixed carbon, calorific value, moisture and vitrinite reflectance. The ASTM classification of coals (Table 1) is based only on rank and uses volatile matter and heating value. Some classifications include grade parameters but most do not include a type parameter. However, the New International Classification (ECE) for codification of higher rank coals includes rank, type and grade parameters (Table 2). It includes vitrinite reflectance, a reflectogram, maceral composition, crucible swelling number, volatile matter, ash, sulphur and gross calorific value (daf). Carpenter (1988), gives an inclusive discussion of coal classification. The reflectogram in the ECE Classification provides a means of distinguishing between single seam coals and blends of coals of different rank. Countries with Gondwanan coals feel the inclusion of inertinite places their coals at a disadvantage.

Class	Group	Fixed Carbon		Volatile Matter		Btu		Agglomerating Character
		≥	<	>	≤	≥	<	
Anthracitic	Semianthracite	86	92	8	14	...	...	Nonagglomerating
Bituminous	Low vol.	78	86	14	22	...	...	⎫
	Med. vol.	69	78	22	31	...	...	⎪ Commonly
	High vol. A	..	69	31	..	14 000	...	⎬ agglomerating
	High vol. B	..	..	..	..	13 000	14 000	⎭
	High vol. C	..	..	..	..	11 500 / 10 500	13 000 / 11 500	Agglomerating
Subbituminous								Nonagglomerating

Table 1. ASTM Classification of Coals by Rank (D388).

A new scientific classification of coal is now being considered by the ECE and is likely to incorporate many of the features being considered by the International Standards Organisation (ISO) and discussed by the International Committee for Coal Petrology (ICCP). It comprehends changes in rank and type of coal and the facies changes from coal through shale and also the fusic to vitric to liptic changes into the oil-rich rock equivalents. The authors' simplified version of this classification is shown in Figure 2.

5 COKING IN A SINGLE COAL PARTICLE

Figure 3 shows graphically the transformation of a coal particle to coke as given by Gray (1987). When an angular particle of coal is heated with restricted access to air, the poorly conducting material loses surface moisture and occluded gases such as water vapour, carbon dioxide, oxygen, nitrogen and methane at temperatures below 200°C. Between 200° and 300°C, the particle becomes slightly rounded. At the first critical temperature level of 350°-450°C, tars and oils evolve and pores develop. The tars and oils or 'Metaplast' migrate through the open subcapillary pores and soften the reactive coal macerals which form an impervious plastic (fluid) mass.

As the temperature increases between 350°-500°C, the gases that are trapped in the plasticized particle expand and the particle swells, forming a coke

Table 2
ECE Codification of Higher Rank Coals[a]

Vitrinite reflectance (Mean Random)		Reflectogram[b]		Maceral Composition				Crucible Swelling Number	
			Standard	Inertinite[c]		Liptinite			
Code	R random, %	Code	Deviation	Code	Vol. %	Code	Vol. %	Code	Number
02	0.2 - 0.29	0	0.1 no gap	0	0 - 9	1	1 - 4	0	0 or 0.5
03	0.3 - 0.39	1	0.1 - 0.2 no gap	1	10 - 19	2	5 - 9	1	1 or 1.5
04	0.4 - 0.49	2	0.2 no gap	2	20 - 29	3	10 - 14	2	2 or 2.5
-	-	3	0.2 1 gaps	-	-	-	-	-	-
49	4.9 - 4.99	4	0.2 2 gaps	8	80 - 89	8	35 - 39	8	8 or 8.5
50	5.0	5	0.2 2 gaps	9	90	9	40	9	9 or 9.5

Volatile Matter[d] (daf)		Ash (dry)		Total Sulfur (dry)		Gross Calorific Value (daf)	
Code	Percent	Code	wt. %	Code	wt. %	Code	MJ/kg
48	48	00	0 - 0.9	00	0.0 - 0.19	24	24 - 24.98
46	46 - 48	01	1 - 1.9	01	0.1 - 0.19	25	25 - 25.98
44	44 - 46	02	2 - 2.9	02	0.2 - 0.29	26	26 - 26.98
-	-	-	-	-	-	-	-
02	2 - 4	20	20 - 20.9	20	2.0 - 2.09	35	35 - 35.98
00	2						

a Classification does not include the lower rank coals: mean random vitrinite reflectance 0.6%, provided the gross calorific value (maf) 24 MJ/kg
b Code number 2 may correspond not only to a blend of medium rank coals, but also to a high rank seam coal

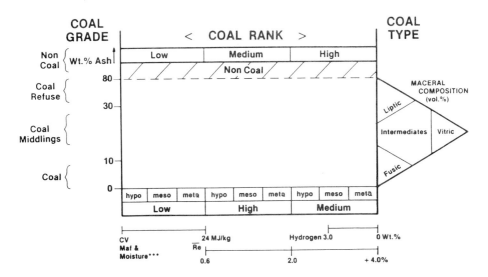

Figure 2. Schematic Diagram of Basic Characteristics (Rank, Type and Grade) for Proposed "International Classification of Coal".

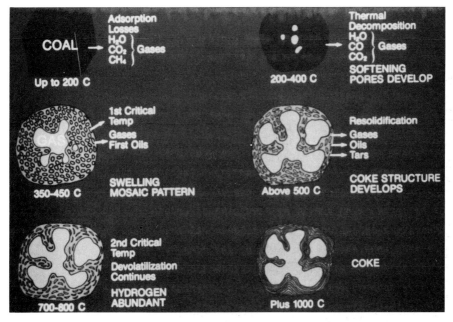

Figure 3. Coking a Single Coal Particle.

'Cenosphere'. At about 425°C, the plasticized reactive macerals of good coking coals form a mesophase which consumes the isotropic parent liquor forming an anisotropic ordered 'mosaic' structure. At 400° to 600°C, additional high molecular weight liquids and gases evolve and, at about 600°C, the plasticized materials solidify forming a semicoke. Lower molecular weight liquids and gases are evolved at 600° to 700°C. From 700° to 1000°C is the second critical temperature where low molecular weight gases cease to be evolved and large amounts of hydrogen are released. Above 700°C, the semi-coke is transformed into coke and the material shrinks to form the final coke by 1000°C.

5.1 Coke cenosphere

Coals above the rank of high volatile bituminous B or C and below the rank of semi-anthracite produce thermally stable metaplast when heated in an inert atmosphere. Gases trapped in the plasticized particles of coals swell to as much as 40 times the coal size to form hollow spheres called cenospheres, as shown in Figure 4 (Gray and Champagne 1988). These hollow spheres were noted in dust explosions in British mines as early as 1910. Sinnatt (1928) experimentally produced cenospheres and described their pore, window and wall structure as well as window inclusions in great detail.

Below a certain size, all of the vapours and gases can escape without forming cenospheres and above a certain size, depending upon coal rank and type,

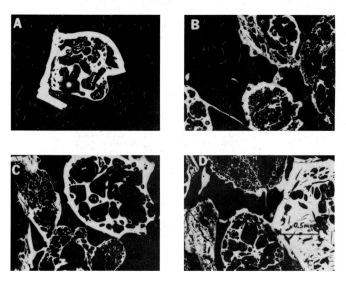

Figure 4. Photographs of Censopheres in Polished Surface Sections from Various Ranks of Coal.

particles swell excessively and form complex structures that only vaguely resemble cenospheres. Any structure that allows gas to escape will impact on the microstructure. When coal is coked in confinement, the cenospheres and their equivalents fuse together to form the coke microstructure.

5.2 Mesophase concept

In the mesophase concept, coking coal is transformed from coal to coke by way of liquid crystal development due to pyrolysis. The development of the anisotropic mosaic structure in coke is attributed to mesophase (liquid crystal) development. This phenomenon is easiest to demonstrate when certain petroleum refinery feedstocks, and/or coal tar pitches (Figure 5) are carbonized. During heating of the parent material, spherical anisotropic liquid crystals (nematic) form and grow and coalesce to form large domains that, with time, form a solid with a mosaic anisotropic pattern. Marsh and Clarke (1986), give a detailed mechanism for the formation of coke structures. Many of the properties of carbons such as electrical resistivity, hardness, resilience and strength can be related to the microstructure. The microstructure can be controlled by the selection of feedstocks and control of the carbonization process.

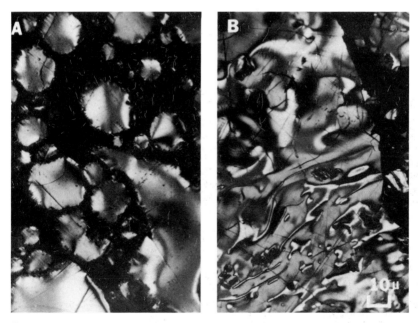

Figure 5. Photomicrographs of A - Mesophase in Heated Coal Tar Pitch and B - Coal Tar Pitch Coke.

5.3 Role of plastic layer in coking

The unique characteristic of coking coals is their ability to soften and become plastic when heated in an inert atmosphere. Coal is transformed to the coke in the plastic layer. This isothermal surface consists of the resolidified coke on the wall or coke side of the oven and the advancing plastic surface in contact with the coal. The plastic layer is a temperature controlled phenomenon. The microscopic appearance of the plastic layer is shown in Figure 6, after Gray and Champagne (1988). The coal first devolatilizes and becomes rounded, then pores develop to mark the plastic layer boundary on the coal side. The pores are swollen or enlarged on the coke side of the plastic layer. The swollen pores shrink as semi-coke forms. The temperature in the plastic layer ranges from about 325° to 600°C and the plastic range is about 50° to 100°C. The mesophase formation was not optically apparent in the samples examined (Figure 6).

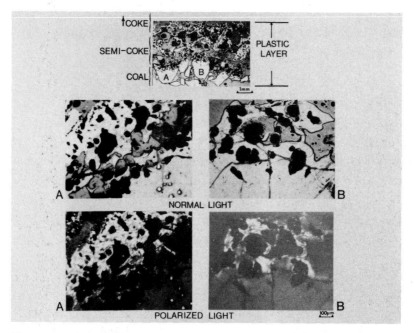

Figure 6. The Macroscopic and Microscopic Appearance of the Plastic Layer from a Pilot Oven Charge.

5.4 Thermal transformation of coal macerals (Anthrathermotics)

Coal macerals behave differently when heated as can be observed when using a microscope heating stage. In the period 1957 to 1960, U.S. Steel conducted thermal studies of coal macerals which they called 'anthrathermotics'. Schapiro et al. (1962) and Schapiro and Gray (1966) found that vitrinite, resinite, exinite, and part of semifusinite and some other inertinite group macerals become plastic when heated and they called this group 'reactives' or coke bond formers. They found that part of the semifusinite and most of the micrinite which was taken to include micrinite, macrinite and inertodetrinite and all of the fusinite was relatively inert when heated and they named this group 'inerts' because they act as fillers in forming the coke. Mineral matter was included as inerts. Some of the thermal transformations of coal macerals are shown in Figure 7.

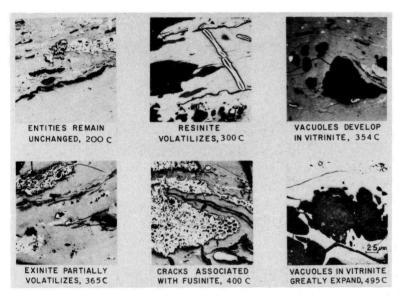

Figure 7. Photomicrographs of the Thermal Transformation of Coal Macerals (Anthrathermotics).

Vitrinite commonly becomes plastic between 330° and 360°C and coarse vitrinite (vitrain) develops elliptical pores, while attrital vitrinite develops fusiform pores (Figure 7). Pores develop first in the resinite cell fillings in telinite. Vitrinite loses its bireflectance near its softening temperature and vitrinites, above 0.85% reflectance, develop a mosaic coke structure at about 425° to 500°C, then solidify and shrink at temperatures above 700°C.

Resinite (terpene) softens and volatilizes almost completely below 200° to 300°C while the telacollinite cell fillings in vitrinite soften between 300° and 350°C, forming pores.

Liptinite or exinite in high volatile coals softens at about 400°C and becomes very fluid at about 410° to 425°C (Figure 8). Much of the liptinite volatilizes to form gas, oils and tars or reacts with vitrinite. Their residues are not distinct in coals with vitrinite reflectance below 0.85%. Liptinite in higher rank coal leaves an anisotropic residue with larger domains than the associated vitrinite. At about 1.35% vitrinite reflectance, the liptinite becomes optically indistinct, first in alkali ash coals and later in acid ash and duller coals.

Semifusinite is transitional between vitrinite and fusinite and displays transitional thermal properties. Semifusinite, associated with vitrinite, fused to vitrinite when heated but isolated semifusinite is not easily wetted. Some semifusinite (not oxy or pyro related) shows anisotropism in the coke.

Micrinite may or may not be inert but it is most commonly abundant in poorly coking coal microlithotypes. Macrinite commonly is semi-reactive, depending upon its reflectance, while inertodetrinite ranges from semi-reactive to inert.

Fusinite is inert in coking and is a centre for crack formation (Figure 8).

Some of the ideas developed from the anthrathermotic studies were incorporated into the U.S. Steel's coke strength prediction system that is based on coal petrography.

6   RHEOLOGICAL PROPERTIES OF COAL

Coking coals have the distinct property of becoming plastic and swelling when heated and a wide variety of tests have been introduced to measure these properties, as described by Brewer (1945). The free-swelling index test is widely used to determine the agglomerating and swelling characteristics of heated coal.

The Gieseler plastometer is commonly used in the U.S.A. and Japan for measuring the plastic properties of heated coal.

In Europe, the Audibert-Arnu dilatometer is commonly used to record the volume changes as a function of time.

7   COAL PETROGRAPHY

Coal petrography has recently developed into a practical tool for selecting and blending coals for coke making. The development of microscopic characterization of coals in terms of their maceral content and maceral maturity has helped in understanding how these factors impact on the carbonization properties of coal.

Schapiro and Gray (1960) published a petrographic classification applicable to coals of all ranks. The essentials of this system are shown in Table 3 and macerals are shown in Figure 8. This system comprehends the coal entity or maceral composition and incorporates a rank or reflectance parameter for each maceral and also includes minerals. The vitrinites are divided into Vitrinite Types or V-Types, each of which represents 0.1% reflectance so a V-Type 8 contains all of the vitrinite with a reflectance of 0.80 to 0.89% in oil.

Table 3
Classification of Coal Macerals or Entities
(After U.S. Steel)

Maceral

Suite	Groups	Coal Macerals & Minerals
Vitrinite	Fusible Vitrinoids	V6 to V19
	Inert Vitrinoids	V0 to V5 and V20 to V70
	Fusible Semifusinoids	SF6 to SF19
Liptinite	Exinoids	E0 to E13
	Resinoids	R0 to R19
Inertinite	Micrinoids	M18 to M70
	Fusinoids	F30 to F70
	Inert Semifusinoids	SF20 to SF29
Mineral	Sulphides	Pyrite, Marcasite
	Carbonates	Calcite, Siderite, etc
	Silicates	Illite, Kaolinite, etc.

The Schapiro-Gray classification is the basis of the ASTM D2796-82 and 2797-80, and 2798-79 for coal pellet preparation, maceral and reflectance determination. This system is widely used in the U.S.A. and the vitrinite reflectance concept is accepted worldwide.

The ideas that are incorporated in the Schapiro-Gray classification served as a prerequisite to the use of petrographic data for predicting coking properties of coal.

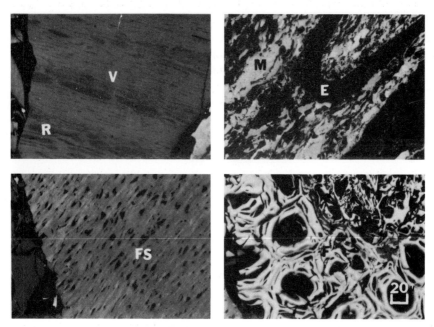

Figure 8. Photomicrographs of Coal Macerals, V = Vitrinite, R = Resinite, E = Exinite, M = Micrinite, SF = Semi-fusinite and F = Fusinite.

8 COKE STRENGTH PREDICTION

A major industrial application of coal petrography is the prediction of coke strength and other properties based on petrographic data. The U.S.A. systems are aimed at predicting coke stability (ASTM D-3402). The prediction systems minimize the amount of empirical coke testing and also permit the evaluation of coal based on limited sample size. The technique also permits a mechanistic approach to blending coals to achieve specific properties in the coke. In the U.S.A., the Schapiro, Gray and Eusner system (1961), which was developed at U.S. Steel about three years earlier has been widely accepted. The Russian system of Ammasov et al. (1957), preceeded the U.S. systems. The Bethlehem Steel group recognizes two vitrinite populations, one of which is pseudovitrinite and is considered partially inert.

The Schapiro-Gray (1960) system uses two types of petrographic data and a volume-percent mineral matter in calculating a predicted stability. Schapiro and Gray, (1960), introduced the concept of V-type, which is the reflectance of vitrinite in tenths. The type parameter is the total content of reactives and inerts. These data are used to calculate Composition Balance Index (CBI), which is the ratio of the actual inerts present to the amount required for optimum coke strength. Thus, a CBI of 2 means the coal has twice the inerts needed for best coke strength. When inerts are added to vitrinite, the strength first increases up to some point where the reactives can no longer assimilate the inerts and then strength decreases (Figure 9a). When inerts are added to different rank vitrinites (different V-types), the same phenomenon occurs but the strength level is different because each V-type not only requires different inert levels for optimum strength but also produces different levels of binder carbon strength (Figure 9b). To comprehend this rank influence, the strength index (SI) or rank index (RI) was introduced. The two ideas can be assimilated and related to empirical strength as shown in Figure 10 and a graphic system of prediction emerges. This system can be used to enhance individual coals in preparation and to select coals for blending to achieve specific levels of strength.

The Schapiro-Gray System is about 30 years old, so that numerous modifications have emerged. Some of the Japanese have followed this system while others have retained vitrinite reflectance as the rank parameter but introduced the log of maximum fluidity from the Gieseler plastometer test as the second parameter. The fluidity reflects coal type changes as well as oxidation, etc. It is important to know the strength of coke that a coal will produce; but, it is also very important to know the volume changes because coke masses must shrink sufficiently to permit easy discharge from the coke ovens.

A scheme for the prediction of volume change during carbonization, based on petrographic data, has been developed (Figure 11).

It is also important to know the carbonization pressures that will be generated by a coal blend, so that coke oven walls are not damaged. A scheme for predicting coking pressures based on coal petrography has also been developed (Figure 12).

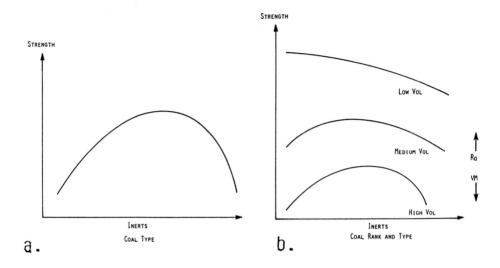

Figure 9a. Change in Coke Strength with Increasing Inerts.
9b. Change in Coke Strength with Rank.

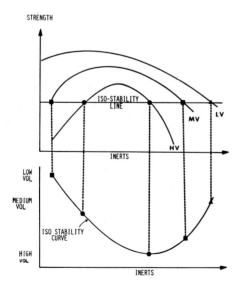

Figure 10. Relation of Coal Types (Inerts) and Rank (Vitrinite Reflectance) to Coke Strength (Stability).

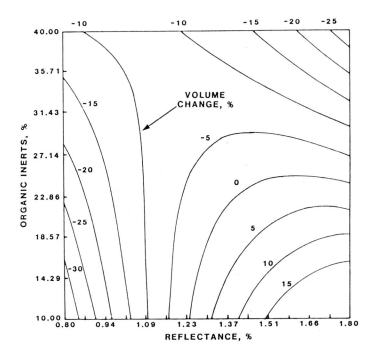

Figure 11. Relation of Coal Rank (Vitrinite Reflectance) and Type (Organic Inerts) to Volume Change During Carbonization.

9 COKE PETROGRAPHY

It has long established that various carbonaceous materials such as petroleum residues, coals of different rank and coal-tar pitches produce coke carbon forms that are distinct. Many techniques have been developed for characterizing coke microstructure. Recently, Gray (1976), Gray and DeVanney (1986), Patrick et al. (1977a, 1977b, 1979) and Marsh (1982) have characterized cokes according to their microtexture.

The U.S. Steel method of coke petrography dates to the period between 1959 and 1962 which followed their anthrathermotic studies. Schapiro and Gray introduced the V-type concept in 1960 and believed that each vitrinoid type ranging from V-7 through V-17 produced a distinct and recognizable carbon form when fresh coal is coked at about 0.9 to 1-inch per hour.

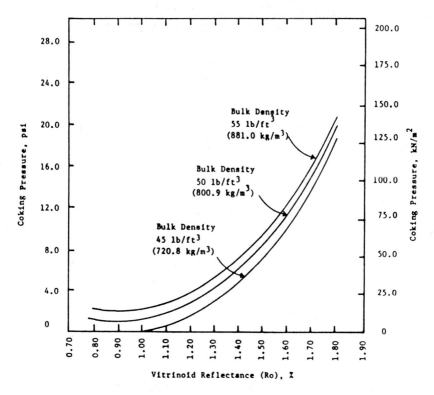

Figure 12. A Scheme for Predicting Coking Pressures based on Vitrinite Reflectance.

Gray's (1976) classification recognized that cokes consist of a binder phase from reactive macerals and a filler phase from inert macerals. The filler phase carbons are essentially isotropic or optically inactive. The binder phase carbons vary with rank in a systematic fashion, the U.S. Steel system is based on these concepts. The idea of the Gray classification (Table 4) was to permit predicting coke carbon forms from a coal analysis or reconstructing the coal composition from a coke analysis.

The binder phase carbons from high volatile coals are from V-types 7 through 11 and range from isotropic through incipient to fine, medium and coarse circular with anisotropic domains of 0.5 to 2.0 microns (Figure 13).

Table 4
A System of Coke Microscopy*
Coke Binder-Phase Carbon Form Classification

Binder Phase	Width (in Microns)	Length (L) to Width (W) Relation	Parent Coal Vitrinoid Type
Isotropic	0.0	None	6, 7
Incipient (anisotropic)	0.5	L = W	8
Circular (anisotropic)			
Fine circular	0.5-1.0	L = W	9
Medium circular	1.0-1.5	L = W	10
Coarse circular	1.5-2.0	L < 2W	11
Lenticular (anisotropic)			
Fine lenticular	1.0-3.0	L $\geq$ 2W, L < 4W	12
Medium lenticular	3.0-8.0	L > 2W, L < 4W	13
Coarse lenticular	8.0-12.0	L > 2W, L $\leq$ 4W	14
Ribbon (anisotropic)			
Fine ribbon	2.0-12.0	L > 4W	15
Medium ribbon	12.0-25.0	L > 4W	16
Coarse ribbon	25.0+	L > 4W	17, 18

Coke Filler-Phase Carbon Form Classification

Filler-Phase	Size (Microns)	Precursors
Organic inerts		
Fine	<50	micrinite, macrinite, inertadetrinite
Coarse	>50	semifusinite, fusinite durain
Miscellaneous inerts		
Oxidized coal (coke)		oxidized and brecciated coal
Brecciated coal (coke)		
Noncoking vitrinite (coke)		vitrinite too high or low in rank
Inorganic inerts		
Fine	<50	mineral matter and coal bone
Coarse	>50	

*Miscellaneous categories including carbon additives, depositional carbons and green and burnt coke may be quantified.

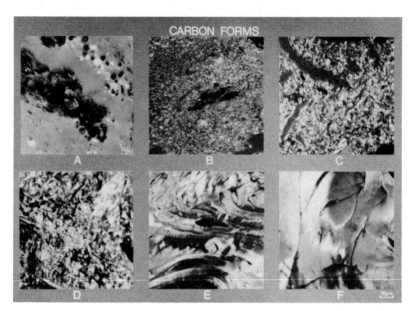

Figure 13. Photomicrographs of Coke Carbon Forms from Various Ranks of Coal. A is isotropic Carbon from Marginal Coking High Volatile Coal, B is Circular Anisotropic Carbon from Fluidity High Volatile Coal, C is Transitional Circular to Lenticular Carbon, D is Lenticular Carbon from Medium Volatile Coal and E and F are Ribbon Anisotropic Carbon from Low Volatile Coal.

The binder carbons from medium volatile coals are lenticular in shape, ranging in widths from 1 to 12.0 microns with length (L) to width (W) ratio of $L \geq 2W$ to $L \leq 4W$, (Table 4). They are produced from V-types 12, 13 and 14 (Figure 13). The binder phase carbons from low-volatile coals have ribbon-like domains with widths of 2 to greater than 25 microns and length to width ratios of $L > 4W$. They are from V-types 15, 16 and 17 plus. The interference colours range from red to blue.

Coal blends can be approximated by point-counting the carbon forms and assigning the carbon forms to the binder phase that produced the coked area upon which the point falls regardless of the phase, be it filler, binder or mineral. The carbon counts are corrected for yield.

Filler phase carbons are counted into various size groups of organics and inorganics. This gives an indication of the types and grades of coals used to make the cokes. This is a separate count from the blend proportioning count. In addition, the carbons from oxidised and brecciated coals and depositional and green and burnt carbons are counted to give additional information on the coals used and the coking conditions. Additive carbons such as breeze, anthracite and petroleum coke are counted as part of the coal blend.

The coke-carbon form analyses can be used to predict strength and reactivity and estimate pressures and operating conditions such as green pushes, delays, etc.

10 COKE PORE AND WALL STRUCTURE

The pore and wall size and distribution in addition to carbon forms are important in characterizing cokes. The range in variability in pore and wall microstructure in cokes from different ranks of coal is illustrated in Figure 14. "A" is from a poorly coking, high volatile coal, "B" is from fluid high volatile coal, "C" is from strongly coking medium volatile coal, "D" is from low volatile rank coal.

High volatile bituminous coals tend to produce the largest amounts of by-products and have the lowest coke yield and make porous and fissured coke. Medium volatile bituminous coals are usually the best coking coals and produce moderate porosities with relatively thick coke walls and their carbonization pressures can be acceptable. Low volatile coals may or may not produce acceptable cokes in terms of strength but they are seldom coked alone since they generate excessive coke-oven wall pressures.

Coke pore and wall size are also affected by the coal size, heating rate and bulk density. A decrease in coal size maximises particle interaction and leads to an homogeneous coke microstructure provided there are not too many fines below the size that will form cenospheres. Increasing heating rate increases porosity and crack formation for most coking coals and is most deleterious for strongly coking medium volatile coals. Increased bulk density decreases porosity and increases strength but does not necessarily increase coke size. It is most effective for marginal coking coals. Increasing the heating rate also increases the size of anisotropic domains in the coke carbon.

Coal fluidity effects pore and wall development. Coals with excessive Gieseler fluidity tend to be more porous with thinner walls while coals with insufficient fluidity tend to be more dense and granular with poorly fused particles. Gieseler fluidity also affects the carbon forms that are developed. Figure 15 shows coke from coals with the same reflectance but with different fluidities.

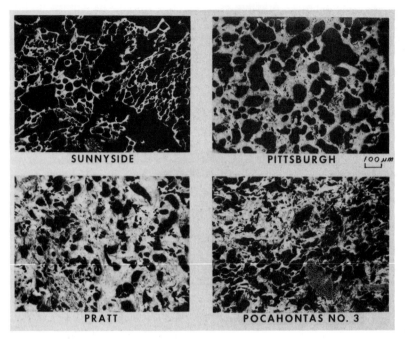

Figure 14. Photomicrographs of Coke Microstructures from Sunnyside (Marginal Coking High Volatile), Pittsburgh (Fluid High Volatile), Pratt (Strongly Coking Medium Volatile) and Pocahontas (Low Volatile).

The occurrence of large amounts of finely disseminated inerts produces larger anisotropic domains in the binder carbon than similar coals with less inerts or coarser inerts. This is because liptinite tends to be associated with the finer-sized organic inerts and differences in the attrital vitrinite that occurs in association with the fine size inerts. Coarse inerts restrict binder phase anisotropic domain size particularly when the inerts occur in sandwich-like layers. Mineral matter, both clays and pyrite, are associated with the development of less anisotropic carbon, particularly in the areas around the mineral concentrations. Coke breeze additions tend to increase coke size by acting as antifissurants. Breeze provides microfissures to relieve stress in the coke structure. The role of breeze changes with its size because the finer sizes act as inerts and enter into the wall structures while the coarser sizes act to increase coke size and decrease gross fissure development. The rank and fluidity of the blend coals must be considered when breeze is added.

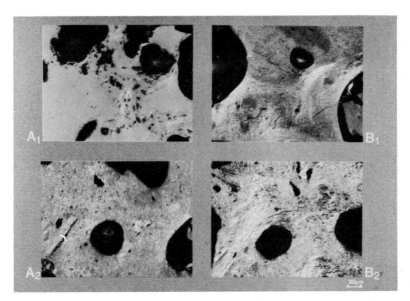

Figure 15. Coke from Coals with the Same Reflectance. A = 1.0 Ro, while A1 has a fluidity of 2,000 ddpm and A2 has 30,000 ddpm. B = 1.15 Ro, while B1 has a fluidity of 1,000 ddpm and B2 is 6,000 ddpm.

Variables in vitrinite distributions within coals and variabilities in maceral distribution result in differences in coke strength between coals with the same vitrinite reflectance and maceral composition.

It is apparent from the above discussion that several factors impact on total and effective coke porosity and wall microstructures and carbon forms. Considerable experience is required to interpret coke microstructures and carbon forms.

11 COAL BLENDS FOR COKE MAKING AND BLAST FURNACE COKE PROPERTIES

11.1 Coal blend

The coal selected to make coke is obviously the most important variable that controls coke properties. The rank and type of coals selected impact on coke strength. Coal chemistry largely determines coke chemistry. The coke yield ash and sulphur can be calculated from coal properties by the following formulae:

1. Coke ash = (Coal ash x 100)/(coke yield)).
2. Coke sulphur = 0.759 x coal sulphur + 0.08.
3. Coke yield = 95.1837 - 0.76107 x VM (dry).

In general, bituminous coals are selected for blending to make blast furnace coke of the highest strength with acceptable reactivity and at a competitive cost. The blend must contract sufficiently for easy removal from the oven and pressure must be acceptable.

Good coal blends for coke production have the following specifications:

Specifications for Coking-Coal Blends

	Good	Average
Ash, wt.%	6.0	8.0
Sulphur, wt.%	0.7	1.0
Ash-Fusion Temperature, °C	1370	1260
Phosphorus, wt.%	0.01	0.03

11.2 Carbonization variables

The coking industry, worldwide, has a symbiotic relationship with the steel industry and coking operations flourish as the steel industry flourishes. Thus, coke-oven operations are governed by the demand for steel.

In general, the selection of the proper blend of coals accounts for about 80% of the coke properties and about 20% is due to the preparation and carbonization variables.

Effect of Coking Rate is the most important carbonization variable relative to oven operations. Coking rate is the rate of travel of the plastic layer. Coke stability generally decreases as the coking rate increases and coke hardness increases. However, increased heating rates can result in higher QI's and hotter tops, more carbon formation in the ovens and, in some cases, increased emissions and increased pressures.

Bulk Density increase usually increases coke strength because coal particle contacts are increased, shrinkage of the charge is decreased and the coke density is increased. Higher bulk densities result in higher coking pressures and less contraction.

Effect of Pulverization is important in blending coals and ensuring the proper size distribution between the coal bond forming macerals and the filler phase. Under the best conditions, the individual coals are pulverized separately and then blended. Coals that are heterogeneous in their distribution of macerals and/or minerals must be crushed finer than less variable coals to assure a more homogeneous distribution of the coke-forming materials.

In general, coke strength increases as coal size decreases if the bulk densities of the charge are the same and there is a limited amount of fines below the size that forms cenospheres.

11.3 Coke strength determinations

ASTM Tumbler Test for Coke   D-3402-75 is commonly used in the U.S.A. to test coke strength (stability and hardness). Tumbled coke is screened at 1 inch (25 mm) and the percentage of plus 1 inch is the stability factor for the coke. The minus 1 inch is screened at $1/4$ inch (6.3 mm) and the plus $1/4$ inch is the hardness factor.

The ISO Micum Test for Coke, widely used in Europe, has been developed from the French Micum drum test where the tumbled coke is screened at 40 mm and 10 mm. The relationship of ASTM stability and hardness to the ISO Micum is as follows:-

$$\text{Stability} = 1.32 \times M_{40} - 43.9$$
$$\text{Hardness} = -0.287 \times M_{10} + 68.6$$

The Japanese standard drum test uses two indices. The plus 15 mm coke after 30 drum revolutions is the DI 30/15 JIS strength index. If the coke is tumbled at 150 revolutions, the DI 150/15 is designated as hardness.

The ASTM Drop Shatter Test, D-3038 measures the resistance to breakage of coke when dropped from a height of 6 feet (1.83 m) on to a steel plate.

11.4 Coking pressure

Coking pressure is the pressure generated in the course of carbonization of coals. It is measured either as the direct pressure on the walls of coke ovens or as gas pressure in the mass during carbonization. Some pressure between the coal particles improves coke strength but excessive pressures damage oven walls. Coking pressures increase as the coal bulk density charged to the ovens and the heating rate increases.

In a previous section, the coking of a single-coal particle is described. The heated coal softens and expands because of trapped gas generated in the particle. However, in the oven, there is a multitude of particles and they are confined so that the actual coal-to-coke transformation takes place in the plastic layer. Soth and Russel (1944) divided the plastic layer or seam into three regions as shown in Figure 16.

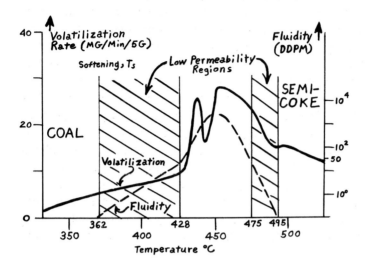

Figure 16. Shows the Volatilization Rate and Gieseler Fluidity for the Plastic Layer during Carbonization.

One curve is the fluidity curve and the other is the volatilization curve. There is a larger temperature interval of low fluidity on the coal side than on the coke side of the plastic layer. Thus, the greatest resistance to gas penetration is on the coal side of the plastic layer. It is estimated that over 80% of the gas exits through the coke on the coke-oven wall side. When the plastic layers meet in the centre of the charge, there is a large low permeability region between two plastic regions. A greater gas pressure may occur when this condition exists.

Gas pressure is easiest to measure and shows promise of being correlated with coking pressure but the plastic layer area must be correlated with the wall area and a sufficient number of probe pressures must be measured.

12 COKE IN THE BLAST FURNACE

Coke in a blast furnace has the following functions:
1. Reducing agent
2. Fuel
3. Supplies permeability

In addition to these functions, coke supports the burden and occupies about 60% or more of the blast furnace volume. It also carburizes the pig iron and reduces the alloying components of the pig iron (Si, P, and Mn).

The daily productivity of a blast furnace depends on the rate at which blast can be supplied without upsetting smooth operation of the furnace and on the amount of coke that must be burned at the tuyeres to produce a ton of iron. The limiting coke rate under normal blast furnace conditions is determined more by gas flow and stock descent rather than by chemistry and thermal requirements. This has to be an oversimplification because carbon can be consumed in the stack under different conditions. The conditions of carbon consumption affect efficiency. The temperature at which the CO_2 gas is formed is the important factor. In general, indirect reduction reactions produce heat and direct reduction absorbs heat. So all of the coke at the tuyeres should be oxidised to CO and oxide reduction above the tuyeres should be by CO. The direct reduction needed is not that of the hearth but direction reduction in the stack. Wustite reduction is the most important. Wustite is spongy and may have an iron coating. The coke may have an ash coating so that contact beween wustite and C of the coke may not be possible. Structures of phases present influence reactions that chemically could take place.

In a blast furnace, the burden is layered as the coke, ore and stone are added. The layers descend in plug flow. The coke, which makes up about 60% of the volume, acts to control permeability and passage of gases from the hot blast. The furnace is hottest in the middle of the charge and cooler along the walls. The upper area in the furnace is the solid or lumpy zone while the middle transitional layer is the cohesive zone where reduction begins. Below the cohesive zone is a moving, loosely packed zone followed by a stagnant, loosely packed zone of coke which grades directly into a deadman or compact coke zone. Final reduction and melting take place below the cohesive zone when the slag separates and floats on the metal in the hearth or bottom of the blast furnace.

Ishikawa et al. (1983) found that an increase in the reactivity and a decrease in after-reaction-strength of coke resulted in increased fine coke in the raceway, expansion of the deadman and a contraction of the raceway depth and active coke zone.

12.1 Coke reactivity

Coke provides support and permeability to the blast furnace burden and also supplies heat and acts as a reductant. It performs similar functions in a smelter, silicon furnaces and lead blast furnaces. It is obvious that coke needs to be strong to support the burden and clean to supply the most heat and reducing gas per unit added without contaminating the product. In addition, it must react with CO_2 to form CO. This reaction depends upon the temperature and CO/CO_2 ratio in the gas. Much of the interest in reactivity of carbon to CO_2 centres around coke consumption in "solution loss" or Boudouard reaction: $C + CO_2 = 2CO$. The Boudouard reaction is controlled by chemical considerations, pore diffusion or boundary layer diffusion depending upon the temperature.

Schapiro and Gray (1963) conducted a review of the relevant literature and studied the reactivity of experimental cokes from single seams or sources and commercial cokes from blends. They found that reactivity decreased as the carbon form (coal rank) increased through medium volatile coal rank, then began to increase with cokes from low volatile coals. Coke porosity tended to follow the same pattern so that reactivity was found to be proportional to pore volume. Other workers found a similar relation but also found that reactivity was related to vitrinite reflectance and inert content of the parent coals. Reactivity of coke decreases with increasing carbon content and increases as the magnesium and calcium levels increase. Reactivity increased as "feeder" pores increased which refers to connecting pores or permeability and coke reactivity decreased as carbonization temperature increased. Taylor and Marsh (1975) discussed the selective or preferential oxidation of structural components and carbon forms and reported isotropic carbon to be most reactive.

Goscinski, Gray and Robinson (1985) studied rank of coals in terms of the geologic and geographic locations, the characteristic chemistry of coal ash and the cumulative effects of rank and ash compositions on reactivity.

In reactivity studies, the following factors are important:-

1. Carbon forms - Rank effect.
2. Coke porosity and permeability - Surface area effect.
3. Mineralogy or ash chemistry - Catalytic effect.
4. Final coke temperature or coke volatile matter.
5. Amount of fractured and original surfaces.
6. Extreme ash content (<5% >15%)
7. High contents of alkaline elements, trace elements, sulphur retention elements, chlorine, oxidised material and organic inerts.

12.2 Petrographic strength predictions for foundry coke

Petrographic data on coal rank and type are widely used in predicting coke stability for blast furnace coke but are less widely used in predicting cupola coke properties. A petrographic prediction technique for cupola coke properties such as shatter strength and size is needed.

12.3 Formed coke

Many formed-coke processes have been tested on a small-scale and about ten have been developed to a more advanced state. The more-advanced processes are FMC BFL, Consol-BNR, EBV, FCL, DKS, HBN, Sapozhnikov, Broken Hill and Nippon Steel. These processes have been developed because where coking coals were in short supply or costly, the product can be made to a more exact size and shape or because the process can be made continuous. These processes utilize both coking and non-coking coal as well as char. The briquetting may be hot or cold and the binder is usually a pitch or some similar material such as fluid coal. Pelletizing is another means of forming the product and the Consol-BNR process makes a nodule in a horizontal kiln.

The FMC coke process was developed by FMC and U.S. Steel Corp. This process is probably the most advanced. It produces a metallurgical coke briquette from both coking and non-coking coals. The product has been tested in five blast furnaces in the U.S. and Wales. In this process, a char is produced and the tar is collected and processed for binder. The char and binder are mixed and briquetted and the product is carbonized. Coking coals are treated to minimize agglomeration.

A very interesting process called 'The U.S. Steel Clean Coke Process' uses a two-stage vertical fluidized bed. Conventionally beneficiated coal is carbonized in a fluidized bed in the presence of a recycle stream of hydrogen-rich gas to remove sulphur and produce a char. The char is pelletized with process-derived tar then cured and calcined to produce a low-sulphur metallurgical coke with hydrogen-rich gas as a by-product. To date, form coke processes have had limited success.

12.4 Pre-heating

Coal is normally charged to coke ovens wet and at ambient temperatures. Since the 1950's, much attention has been given to coal pre-heating. Some cited advantages of pre-heated charging are:-

1. Improved coke quality
2. Increased oven and battery through-put
3. Reduced oven brickwork damage
4. Higher thermal efficiency.

A number of pre-heating systems have been developed. In these systems, poor coking coals can be rendered better coking by pre-heating. This may be due to modified plastic properties or possibly due to a permeable plastic layer which allows tar and other pyrolysis products to react with the coal. Pre-heating does not greatly improve good coking coals.

Pre-heated charges have low bulk densities, risk of fire and with larry car charging, emissions into the atmosphere may be a problem.

12.5 Co-carbonization

Coke quality benefits have been attributed to the co-carbonization of coal and a variety of carbonaceous additives. This is important where good coking coals are lacking or too expensive and where the coals available are high in inerts or too low in rank for coking. In addition, there has been a reported disproportionate increase in by-product yields, particularly the light oils. The prime function of additives is to reduce development of cross-links during heating so large aromatic structures can develop. Additives reduce the viscosity of the plastic phase and broaden the plastic range. Petroleum-based additives are the most available but are relatively ineffective, are costly to reform and are commonly high in sulphur. The most effective additives should be high-boiling, highly aromatic substances and should contain functional groups and mobile hydrogen. Solvent Refined Coal (SRC) has been found to be very effective and reduces the reactivity of the resultant coke.

Enhanced by-product yield has promoted research into co-carbonization additives. Additives are known to increase yields of tar, light oil and gas.

References

Ammasov, I.I., Eremin, I.V., Sukhenko, S.F. and Oshurkova, L.S., (1957). Calculation of coking changes on basis of petrographic characteristics of coals. Koks i Khimiyo, **12**, 9.

Brewer, R.E. (1945). Plastic, Agglutenating, Agglomeration and Swelling Properties of Coals in H.H. Lowry Ed., Utilization of Coal, Vol. **1**, pp. 170-175.

Brooks, J.D. and Taylor, G.H. (1968) The formation of some graphitizing carbons. The Chemistry and Physics of Carbon, Ed. Walker, P.L. Jr. Marcel Dekker Inc., N.Y. **4**, pp. 243-285.

Carpenter, A.M. (1988) Coal Classification IEA Coal Research, No. IEA CR/12 London, U.K. pp. 1-104.

Chermin, H.A.G. and van Krevelen, D.W. (1957) A mathematical model of coal pyrolysis, Fuel 36, 85.

Finney, C.A. and Mitchell, J. (1961). A history of the coking industry in the United States. Journal of Metals, April - August, pp. 5-31.

Fischer, F. and Gluud, W. (1916). Die Ergiebigkeit der Kohlenextraktion mit Benzol. Ber. Deut. Chem. Ges. 49, Part 1, pp. 1460-1468.

Fitzgerald, D. (1956) The kinetics of coal carbonizations in the plastic state. Trans. Far. Soc. 52, 362.

Fremy, E. (1861). Recherches chimique sur les combustibles mineraux. Compt. Rend. 52, Part 1, pp. 114-118.

Given, P.H. (1960). The distribution of hydrogen in coals and its relation to coal structure. Fuel 39, 147.

Goscinski, J.S., Gray, R.J. and Robinson, J.W. (1985). A review of American coal quality and after strength of cokes. The Journal of Coal Quality 4(1) pp. 35-43, and 4(2), Part II, pp. 21-29.

Gray, R.J. (1976). A system of coke petrography, Illinois Mining Institute Proceedings, pp. 20-47.

Gray, R.J. (1987). Theory of Carbonization of Coal. The First International Meeting on Coal and Coke Applied to Coke Making. Brazilian Society of Metals, ABM, August 9-15, pp. 551-579.

Gray, R.J. and Champagne, P.E. (1988) Petrographic characteristics impacting the coal to coke transformation. Presented at 47th Ironmaking Conference of the Iron and Steel Society, AIME, Toronto, Canada, April 17-20.

Gray, R.J. and DeVanney, K.F. (1986). Coke carbon forms. Microscopic Classification and Industrial Applications. International Journal of Coal Geology, Elsevier Science Publishers B.V. Amsterdam. 6, pp. 277-297.

Ishikawa, Y., Kase, M., Abe, Y., Ono, K., Sugata, M. and Nishi, T. (1983). Influence of post-reaction strength of coke on blast furnace. Ironmaking Proceedings, Vol. 42, Atlanta, Georgia.

Kirov, N.Y. and Stephens, J.N. (1967). Physical aspects of coal carbonization. The University of New South Wales. Sydney, Australia, Chapters 1-3.

Kreulen, D.J.W. (1948). Elements of Coal Chemistry. Rotterdam, p.170.

Krevelen, D.W. van (1950). Graphical statistical method for the study of structure and reaction processes of coal. Fuel **29**, 269.

Krevelen, D.W. van and Schuyer, J.(1957). Coal Science, Elsevier, Amsterdam.

Kirov, N.Y. and Stephens, J.N. (1967). Physical Aspects of Coal Carbonization. The University of New South Wales, Sydney, Australia, Chapters 1-3.

Marsh, H. (1982). Metallurgical Coke: Formation, structure and properties. (AIME) Ironmaking Proceedings. **41**, pp. 2-11.

Marsh, H, and Clarke, D.E. (1986). Mechanisms of formation of structure within metallurgical coke and its effect on coke properties. Erdol und Kohle. **39**, (3) 113.

Marsh, H. and Menendez, R. (1988). Carbons from pyrolysis of pitches, coals and their blends, Fuel Proc. Tech. **20**, 269.

Marsh, H, and Neavel, R.C. (1980). A common stage in mechanisms of coal liquefaction and carbonization of coal blends for coke making. Fuel **59**, 511.

Marsh, H. and Taylor, D.W. (1975). Carbonization gasification in the Boudouard reaction. Fuel **54**, 218.

Marcilly, C.de (1862). De l'action des dissoluants sur la houille. Ann. Chim. et Phys. **66**, pp. 167-171.

McGannon, H.E. Ed. (1970). The Making, Shaping and Treating of Steel, Ninth Edition, Herbick and Held, Pittsburgh, PA.

Neavel, R.C. (1976) Coal plasticity mechanism: inferences from liquefaction studies, Proceedings of the Coal Agglomeration and Conversion Symposium, Sponsored by W. Va. Geological Survey and Coal Research Bureau, W. Va. University, pp. 121-133.

Oele, A.P., Waterman, H.I., Gredkoop, M.L. and Krevelen, D.W. van. (1951). Extractive disintegration of bituminous coals. Fuel **30**, 169.

Patrick, J.W., Reynolds, M.J. and Shaw, F.H. (1979). Optical anisotropy of carbonized coking and caking coals. Fuel **58**, 69.

Patrick, J.W., Shaw, F.H. and Willmers, R.R. (1977a). Microscopic examination of polished coke surfaces etched by ionic bombardment. Fuel **56**, 81.

Patrick, J.W., Sims, M.J. and Stacey, A.E. (1977b). Quantitative characterization of the texture of coke. Journal Microscopy. **109**, 137.

Ramdohr, P. (1928). Mikroskopische Beobachtungen an Graphiten und Koksen. Arch Eisenhuttenwesen **1**, pp. 669-672.

Schapiro, N. and Gray, R.J. (1960). Petrographic classification applicable to coals of all ranks. Proceedings of the Illinois Mining Institute. 68th year, pp. 83-97.

Schapiro, N. and Gray, R.J. (1963). Relation of coke structure to reactivity, Blast Furnace and Steel Plant. April, pp. 273-280.

Schapiro, N. and Gray, R.J. (1966), Petrographic composition and coking characteristics of Sunnyside coal from Utah, A Guidebook for the Geological Society of America and Associated Societies, Utah Geology and Mining Survey, Salt Lake City, Utah, Bulletin 80, pp. 55-79.

Schapiro, N. Gray, R.J. and Eusner, G.R. (1961). Recent developments of coal petrography. AIME Blast Furnace Coke Oven and Raw Materials Proceedings. Vol. **20**. pp. 89-112.

Schapiro, N., Peters, J.T. and Gray, R.J. (1962). Know your coal. Transactions of AIME, March, 1-6.

Sinnatt, F.S. (1928). Some fundamentals of the carbonization of coking coals: the formation of cenospheres. Proceedings of the Second International Conference on Bituminous Coal, Pittsburgh, PA., U.S.A. November 19-24, pp. 560-585.

Soth, G.C. and Russel C.C. (1944). Sources of pressure occurring during carbonization of coal. Transactions AIME Vol. **157**, Coal Division.

Stach, E. (1982) Stach's Textbook of Coal Petrology, Gebruder Borntraeger, Berlin.

Wiser, W.H. (1968). A kinetic comparison of coal pyrolysis and coal dissolution. Fuel **47**, 475.

B C.

RETURN CHEMISTRY LIBRARY 2340
TO → 100 Hildebrand Hall   642-3753

B.C.

OCT 8 1990